Hardware Based Packet Classification for High Speed Internet Routers

T0205683

Chad R. Meiners • Alex X. Liu • Eric Torng

Hardware Based Packet Classification for High Speed Internet Routers

 Springer

Chad R. Meiners
Michigan State University
Department of Computer Science
and Engineering
Engineering Building 3115
East Lansing, MI 48824-1226
USA
meinersc@cse.msu.edu

Alex X. Liu
Michigan State University
Department of Computer Science
and Engineering
Engineering Building 3115
East Lansing, MI 48824-1226
USA
alexliu@cse.msu.edu

Eric Torng
Michigan State University
Department of Computer Science
and Engineering
Engineering Building 3115
East Lansing, MI 48824-1226
USA
torng@msu.edu

This material is based upon work supported by the National Science Foundation under
Grant No. 0916044. Any opinions, findings, and conclusions or recommendations
expressed in this material are those of the authors and do not necessarily reflect the views of
the National Science Foundation.

ISBN 978-1-4899-9954-2   ISBN 978-1-4419-6700-8 (eBook)
DOI 10.1007/978-1-4419-6700-8
Springer New York Dordrecht Heidelberg London

Printed on acid-free paper

Springer is part of Springer Science+Business Media (www.springer.com)

This book is dedicated to chance encounters at the correct time.
– Chad Meiners

Dedicated with love and respect
to my parents Shuxiang Wang and Yuhai Liu (God rest his soul),
to Huibo Heidi Ma
to my twin sons Max Boyang and Louis Boyang,
to whom I owe all that I am and all that I have accomplished.
– Alex X. Liu

Dedicated to Pat and Annika.
– Eric Torng

Contents

List of Figures

List of Tables

Chapter 1
Introduction

Packet classification, which is widely used on the Internet, is the core mechanism that enables routers to perform many networking services such as firewall packet filtering, virtual private networks (VPNs), network address translation (NAT), quality of service (QoS), load balancing, traffic accounting and monitoring, differentiated services (Diffserv), etc. As more services are deployed on the Internet, packet classification grows in demand and importance.

The function of a packet classification system is to map each packet to a decision (i.e.,, action) according to a sequence (i.e.,, ordered list) of rules, which is called a packet classifier. Each rule in a packet classifier has a predicate over some packet header fields and a decision to be performed upon the packets that match the predicate. To resolve possible conflicts among rules in a classifier, the decision for each packet is the decision of the first (i.e., highest priority) rule that the packet matches. Table 1.1 shows an example packet classifier of two rules. The format of these rules is based upon the format used in Access Control Lists on Cisco routers.

Rule	Source IP	Destination IP	Source Port	Destination Port	Protocol	Action
r_1	1.2.3.0/24	192.168.0.1	[1,65534]	[1,65534]	TCP	accept
r_2	*	*	*	*	*	discard

Table 1.1 An example packet classifier

Rule	Source IP	Destination IP	Source Port	Destination Port	Protocol	Action
r_1	1.2.3.0/24	192.168.0.1	0	*	*	discard
r_2	1.2.3.0/24	192.168.0.1	65535	*	*	discard
r_3	1.2.3.0/24	192.168.0.1	*	0	*	discard
r_4	1.2.3.0/24	192.168.0.1	*	65535	*	discard
r_5	1.2.3.0/24	192.168.0.1	[0,65535]	[0,65535]	TCP	accept
r_6	*	*	*	*	*	discard

Table 1.2 TCAM Razor output for the example packet classifier in Table 1.1

C. R. Meiners et al., *Hardware Based Packet Classification for High Speed Internet Routers,*   1
DOI 10.1007/978-1-4419-6700-8_1, © Springer Science+Business Media, LLC 2010

1.1 Motivation

To classify the never-ending supply of packets at wire speed, Ternary Content Addressable Memories (TCAMs) have become the de facto standard for high-speed routers on the Internet [Lakshminarayanan et al(2005)Lakshminarayanan, Rangarajan, and Venkatachary]. A TCAM is a memory chip where each entry can store a packet classification rule that is encoded in ternary format. Given a packet, the TCAM hardware can compare the packet with all stored rules in parallel and then return the decision of the first rule that the packet matches. Thus, it takes $O(1)$ time to find the decision for any given packet. In 2003, most packet classification devices shipped were TCAM-based. More than 6 million TCAM devices were deployed worldwide in 2004.

A traditional random access memory chip receives an address and returns the content of the memory at that address. A TCAM chip works in a reverse manner: it receives content and returns the address of the *first* entry where the content lies in the TCAM in constant time (i.e.,, a few CPU cycles). Exploiting this hardware feature, TCAM-based packet classifiers store a rule in each entry as an array of 0's, 1's, or *'s (*don't-care* values). A packet header (i.e.,, a search key) matches an entry if and only if their corresponding 0's and 1's match. Given a search key to a TCAM, the hardware circuits compare the key with all its occupied entries in parallel and return the index (or the content, depending on the chip architecture and configuration,) of the first matching entry.

Despite their high speed, TCAMs have their own limitations with respect to packet classification.

- **Range expansion:** TCAMs can only store rules that are encoded in ternary format. In a typical packet classification rule, source IP address, destination IP address, and protocol type are specified in prefix format, which can be directly stored in TCAMs, but source and destination port numbers are specified in ranges (i.e.,, integer intervals), which need to be converted to one or more prefixes before being stored in TCAMs. This can lead to a significant increase in the number of TCAM entries needed to encode a rule. For example, 30 prefixes are needed to represent the single range $[1, 65534]$, so $30 \times 30 = 900$ TCAM entries are required to represent the single rule r_1 in Table 1.1.
- **Low capacity:** TCAMs have limited capacity. The largest TCAM chip available on the market has 18Mb while 2Mb and 1Mb chips are most popular. Given that each TCAM entry has 144 bits and a packet classification rule may have a worst expansion factor of 900, it is possible that an 18Mb TCAM chip cannot store all the required entries for a modest packet classifier of only 139 rules. While the worst case may not happen in reality, this is certainly an alarming issue. Furthermore, TCAM capacity is not expected to increase dramatically in the near future due to other limitations that we will discuss next.
- **High power consumption and heat generation:** TCAM chips consume large amounts of power and generate large amounts of heat. For example, a 1Mb TCAM chip consumes 15-30 watts of power. Power consumption together with

the consequent heat generation is a serious problem for core routers and other networking devices.

- **Large board space occupation:** TCAMs occupy much more board space than SRAMs. For networking devices such as routers, area efficiency of the circuit board is a critical issue.
- **High hardware cost** TCAMs are expensive. For example, a 1Mb TCAM chip costs about 200 ~ 250 U.S. dollars. TCAM cost is a significant fraction of router cost.

1.2 Contribution

This work describes two methods of addressing packet classification and the related TCAM based issues: *equivalent transformation techniques* and *new architectural approaches*.

Equivalent transformation techniques seek to find semantically equivalent but more efficient classifiers. Three methods of equivalent transformation are TCAM Razor, Bit Weaving, and All-Match Redundancy Removal. TCAM Razor decomposes a multi-field problem into a series of single-field problems; these problems are solved optimally and then recomposed into a greedy multi-field solution. Bit Weaving differs from TCAM Razor in that it treat the multi-field rules as rules with a single field. This new classifier is analyzed for rule adjencies to produce a smaller list of single field rules, which are then converted back into multi-field rules. In contrast, All-match Redundancy Removal identifies a maximal set of rules that can be removed from a packet classifier without changing the packet classifier's semantics.

New architectural approaches seek to modify how the TCAM based packet classifiers operate in order to improve efficiency. We propose two approaches: sequential decomposition and topological transformation. Sequential decomposition decomposes a single d-field packet classification TCAM lookup into a sequence of d 1-field TCAM lookups. Topological transformations provide methods to translate the domain of each packet field into a more efficient representation. Both techniques allow for the efficient utilization of TCAM space. These techniques mitigate the effects of range expansion; however, they also have the unique advantage that they find optimizations beyond range expansion. This advantage allows for sublinear compression.

Chapter 2
Background

We now formally define the concepts of fields, packets, and packet classifiers. A *field* F_i is a variable of finite length (i.e.,, of a finite number of bits). The domain of field F_i of w bits, denoted $D(F_i)$, is $[0, 2^w - 1]$. A *packet* over the d fields $F_1, \cdots, F_d$ is a d-tuple $(p_1, \cdots, p_d)$ where each p_i ($1 \le i \le d$) is an element of $D(F_i)$. Packet classifiers usually check the following five fields: source IP address, destination IP address, source port number, destination port number, and protocol type. The lengths of these packet fields are 32, 32, 16, 16, and 8, respectively. We use Σ to denote the set of all packets over fields $F_1, \cdots, F_d$. It follows that Σ is a finite set and $|\Sigma| = |D(F_1)| \times \cdots \times |D(F_d)|$, where $|\Sigma|$ denotes the number of elements in set Σ and $|D(F_i)|$ denotes the number of elements in set $D(F_i)$.

A *rule* has the form $\langle predicate \rangle \to \langle decision \rangle$. A $\langle predicate \rangle$ defines a set of packets over the fields F_1 through F_d, and is specified as $F_1 \in S_1 \wedge \cdots \wedge F_d \in S_d$ where each S_i is a subset of $D(F_i)$ and is specified as either a prefix or a nonnegative integer interval. A *prefix* $\{0,1\}^k\{*\}^{w-k}$ with k leading 0s or 1s for a packet field of length w denotes the integer interval $[\{0,1\}^k\{0\}^{w-k}, \{0,1\}^k\{1\}^{w-k}]$. For example, prefix 01** denotes the interval $[0100, 0111]$. A rule $F_1 \in S_1 \wedge \cdots \wedge F_d \in S_d \to \langle decision \rangle$ is a *prefix rule* if and only if each S_i is represented as a prefix.

A packet matches a rule if and only if the packet matches the predicate of the rule. A packet $(p_1, \cdots, p_d)$ *matches* a predicate $F_1 \in S_1 \wedge \cdots \wedge F_d \in S_d$ if and only if the condition $p_1 \in S_1 \wedge \cdots \wedge p_d \in S_d$ holds. We use DS to denote the set of possible values that $\langle decision \rangle$ can be. Typical elements of DS include accept, discard, accept with logging, and discard with logging.

A sequence of rules $\langle r_1, \cdots, r_n \rangle$ is *complete* if and only if for any packet p, there is at least one rule in the sequence that p matches. To ensure that a sequence of rules is complete and thus a packet classifier, the predicate of the last rule is usually specified as $F_1 \in D(F_1) \wedge \cdots F_d \in \wedge D(F_d)$. A *packet classifier* $\mathbb{C}$ is a sequence of rules that is complete. The size of $\mathbb{C}$, denoted $|\mathbb{C}|$, is the number of rules in $\mathbb{C}$. A packet classifier $\mathbb{C}$ is a *prefix packet classifier* if and only if every rule in $\mathbb{C}$ is a prefix rule. A classifier with d fields is called a d-dimensional packet classifier.

Two rules in a packet classifier may *overlap*; that is, a single packet may match both rules. Furthermore, two rules in a packet classifier may *conflict*; that is, the two

C. R. Meiners et al., *Hardware Based Packet Classification for High Speed Internet Routers*,   5
DOI 10.1007/978-1-4419-6700-8_2, © Springer Science+Business Media, LLC 2010

rules not only overlap but also have different decisions. Packet classifiers typically resolve such conflicts by employing a first-match resolution strategy where the decision for a packet p is the decision of the first (i.e., highest priority) rule that p matches in $\mathbb{C}$. The decision that packet classifier $\mathbb{C}$ makes for packet p is denoted $\mathbb{C}(p)$.

We can think of a packet classifier $\mathbb{C}$ as defining a many-to-one mapping function from Σ to DS. Two packet classifiers $\mathbb{C}_1$ and $\mathbb{C}_2$ are *equivalent*, denoted $\mathbb{C}_1 \equiv \mathbb{C}_2$, if and only if they define the same mapping function from Σ to DS; that is, for any packet $p \in \Sigma$, we have $\mathbb{C}_1(p) = \mathbb{C}_2(p)$. A rule is *redundant* in a classifier if and only if removing the rule does not change the semantics of the classifier. Furthermore, we define the equivalence relation that classifier $\mathbb{C}$ defines on each field domain and the resulting equivalence classes. We use the notation Σ_{-i} to denote the set of all $(d-1)$-tuple packets over the fields $(F_1, \cdots, F_{i-1}, F_{i+1}, \cdots, F_d)$ and p_{-i} to denote an element of Σ_{-i}. Then we use $\mathbb{C}(p_i, p_{-i})$ to denote the decision that packet classifier $\mathbb{C}$ makes for the packet p that is formed by combining $p_i \in D(F_i)$ and p_{-i}.

Definition 2.1 (Equivalence Class). Given a packet classifier $\mathbb{C}$ over fields $F_1, \cdots,$ F_d, we say that $x, y \in D(F_i)$ for $1 \le i \le d$ are *equivalent* with respect to $\mathbb{C}$ if and only if $\mathbb{C}(x, p_{-i}) = \mathbb{C}(y, p_{-i})$ for any $p_{-i} \in \Sigma_{-i}$. It follows that $\mathbb{C}$ partitions $D(F_i)$ into *equivalence classes*. We use the notation $\mathbb{C}\{x\}$ to denote the equivalence class that x belongs to as defined by classifier $\mathbb{C}$.

In a typical packet classifier rule, the fields of source IP, destination IP, and protocol type are specified in prefix format, which can be directly stored in TCAMs; however, the remaining two fields of source port and destination port are specified as ranges (i.e.,, non-negative integer intervals), which are typically converted to prefixes before being stored in TCAMs. This leads to *range expansion*, the process of converting a non-prefix rule to prefix rules. In range expansion, each field of a rule is first expanded separately. The goal is to find a minimum set of prefixes such that the union of the prefixes corresponds to the range (see Algorithm 1). For example, if one 3-bit field of a rule is the range $[1,6]$, a corresponding minimum set of prefixes would be 001, 01*, 10*, 110. The worst-case range expansion of a w-bit range results in a set containing $2w - 2$ prefixes [Gupta and McKeown(2001)]. The next step is to compute the cross product of the set of prefixes for each field, resulting in a potentially large number of prefix rules.

2.1 Firewall decision diagrams

A crucial data structure required for this work is the Firewall Decision Diagram (FDD) [Gouda and Liu(2004)]. A *Firewall Decision Diagram* (FDD) with a decision set DS and over fields $F_1, \cdots, F_d$ is an acyclic and directed graph that has the following five properties: (1) There is exactly one node that has no incoming edges. This node is called the *root*. The nodes that have no outgoing edges are called *terminal* nodes. (2) Each node v has a label, denoted $F(v)$, such that

Input: An interval *Interval* = (a,b) and a prefix aligned interval *test* = (c,d) s.t.
   a,b,c,d ∈ ℕ.
Output: A list of prefix aligned ranges.

1 Let $i = (e,f)$ be the intersection of *Interval* and *test* ;
2 **if** *i is empty* **then**
3   **return** an empty list ;
4 **else**
5   **if** $i = test$ **then**
6   **return** a list that contains only *test* ;
7   **else**
8   Split *test* into two prefix intervals *low* and *high* ;
9   **return** the concatenation of GetPrefixes(*Interval*,low)) and
              GetPrefixes(*Interval*,high) ;
10   **end**
11 **end**

Algorithm 1: GetPrefixes(Interval, test = $(0, 2^{32} - 1)$

$$F(v) \in \begin{cases} \{F_1, \cdots, F_d\} & \text{if } v \text{ is a nonterminal node,} \\ DS & \text{if } v \text{ is a terminal node.} \end{cases}$$

(3) Each edge $e{:}u \to v$ is labeled with a nonempty set of integers, denoted $I(e)$, where $I(e)$ is a subset of the domain of u's label (i.e., $I(e) \subseteq D(F(u))$). (4) A directed path from the root to a terminal node is called a *decision path*. No two nodes on a decision path have the same label. (5) The set of all outgoing edges of a node v, denoted $E(v)$, satisfies the following two conditions: (i) *Consistency*: $I(e) \cap I(e') = \emptyset$ for any two distinct edges e and e' in $E(v)$. (ii) *Completeness*: $\bigcup_{e \in E(v)} I(e) = D(F(v))$.

We define a *full-length ordered FDD* as an FDD where in each decision path all fields appear exactly once and in the same order. For ease of presentation, we use the term "FDD" to mean "full-length ordered FDD" if not otherwise specified. Given a packet classifier ℂ, the FDD construction algorithm in [Liu and Gouda(2004)] can convert it to an equivalent full-length ordered FDD f. Figure 2.1(a) contains a sample classifier, and Figure 2.1(b) shows the resultant FDD from the construction process shown by Agorithms 2 and 3.

Input: A packet classifier $f : \langle r_1, r_2, \cdots, r_n \rangle$
Output: A t for packet classifier f

1 Build a path from rule r_1. Let v denote the root. The label of the terminal node is $\langle 1 \rangle$. ;
2 **for** $i = \{2, \ldots, n\} \in C$ **do**
3   APPEND($v, r_i, 1, i$) ;
4 **end**

Algorithm 2: FDD Construction Algorithm

After an FDD f is constructed, we can reduce f's size by merging isomorphic subgraphs. A full-length ordered FDD f is *reduced* if and only if it satisfies the

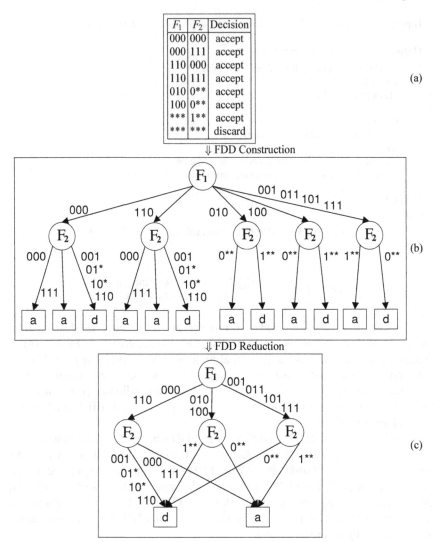

Fig. 2.1 Illustration of FDD construction

following two conditions: (1) no two nodes in f are isomorphic; (2) no two nodes have more than one edge between them. Two nodes v and v' in an FDD are *isomorphic* if and only if v and v' satisfy one of the following two conditions: (1) both v and v' are terminal nodes with identical labels; (2) both v and v' are nonterminal nodes and there is a one-to-one correspondence between the outgoing edges of v and the outgoing edges of v' such that every pair of corresponding edges have identical labels and they both point to the same node. A reduced FDD is essentially a canoni-

Input: A vertex v, a rule $(F_1 \in S_1) \wedge \cdots \wedge (F_d \in S_d) \rightarrow \langle dec \rangle$, a depth m, and a rule number i.
Output: v includes the rule $(F_1 \in S_1) \wedge \cdots \wedge (F_d \in S_d) \rightarrow \langle dec \rangle$ in the FDD

```
        /*  F(v) = F_m  and  E(v) = {e_1,···,e_k}                                      */
 1  if m = d+1 then
 2      Make i's decision of v's label if the decision is not already defined. ;
 3      return
 4  end
 5  else if (S_m − (I(e_1)∪···∪I(e_k))) ≠ ∅ then
 6      Add an outgoing edge e_{k+1} with label S_m − (I(e_1)∪···∪I(e_k)) to v ;
 7      Build a decision path from (F_{m+1} ∈ S_{m+1}) ∧ ··· ∧ (F_d ∈ S_d) → ⟨dec⟩, and make e_{k+1}
        point to the first node in this path ;
 8      Add i to the end of the label of the terminal node of this decision path ;
 9  end
10  for j := 1 to k do
11      if I(e_j) ⊆ S_m then
12          APPEND(e_j's target,(F_1 ∈ S_1) ∧ ··· ∧ (F_d ∈ S_d) → ⟨dec⟩,m+1,i);
13      end
14      Add one outgoing edge e to v, and label e with I(e_j)∩S_m;
15      Make a copy of the subgraph rooted at the target node of e_j, and make e points to the
        root of the copy ;
16      Replace the label of e_j by I(e_j) − S_m ;
17      APPEND(e's target,(F_1 ∈ S_1) ∧ ··· ∧ (F_d ∈ S_d) → ⟨dec⟩,m+1,i) ;
18  end
19  return
```

Algorithm 3: APPEND

cal representation for packet classifiers. Figure 2.1(c) shows the reduced FDD from Figure 2.1(b).

2.2 One-Dimensional Classifier Minimization

The special problem of *weighted* one-field TCAM minimization is used as a building block for multi-dimensional TCAM minimization. Given a one-field packet classifier f of n prefix rules $\langle r_1, r_2, \cdots, r_n \rangle$, where $\{Decision(r_1), Decision(r_2), \cdots, Decision(r_n)\} = \{d_1, d_2, \cdots, d_z\}$ and each decision d_i is associated with a cost $Cost(d_i)$ (for $1 \leq i \leq z$), we define the cost of packet classifier f as follows:

$$Cost(f) = \sum_{i=1}^{n} Cost(Decision(r_i))$$

Based upon the above definition, the problem of weighted one-dimensional TCAM minimization is stated as follows.

Definition 2.2. Weighted One-Dimensional Prefix Minimization Problem Given a one-field packet classifier f_1 where each decision is associated with a cost, find a prefix packet classifier $f_2 \in \{f_1\}$ such that for any prefix packet classifier $f \in \{f_1\}$, the condition $Cost(f_2) \leq Cost(f)$ holds.

The problem of one-dimensional prefix minimization (with uniform cost) has been studied in [Draves et al(1999)Draves, King, Venkatachary, and Zill, Suri et al(2003)Suri, Sandholm, and Warkhede] in the context of compressing routing tables. I generalize the dynamic programming solution in [Suri et al(2003)Suri, Sandholm, and Warkhede] to solve the weighted one-dimensional TCAM minimization. There are three key observations:

1. For any one-dimensional packet classifier f on $\{*\}^w$, we can always change the predicate of the last rule to be $\{*\}^w$ without changing the semantics of the packet classifier. This follows from the completeness property of packet classifiers.
2. Consider any one-dimensional packet classifier f on $\{*\}^w$. Let f' be f appended with rule $\{*\}^w \to d$, where d can be any decision. The observation is that $f \equiv f'$. This is because the new rule is redundant in f' since f must be complete. A rule in a packet classifier is *redundant* if and only if removing the rule from the packet classifier does not change the semantics of the packet classifier.
3. For any prefix $\mathscr{P} \in \{0,1\}^k\{*\}^{w-k}$ ($0 \leq k \leq w$), one and only one of the following conditions holds:

   a. $\mathscr{P} \in \{0,1\}^k 0\{*\}^{w-k-1}$,
   b. $\mathscr{P} \in \{0,1\}^k 1\{*\}^{w-k-1}$,
   c. $\mathscr{P} = \{0,1\}^k\{*\}^{w-k}$.

   This property allows us to divide a problem of $\{0,1\}^k\{*\}^{w-k}$ into two sub-problems: $\{0,1\}^k 0\{*\}^{w-k-1}$, and $\{0,1\}^k 1\{*\}^{w-k-1}$. This divide-and-conquer strategy can be applied recursively.

We formulate an optimal dynamic programming solution to the weighted one-dimensional TCAM minimization problem.

Let $\mathscr{P}$ denote a prefix $\{0,1\}^k\{*\}^{w-k}$. We use $\underline{\mathscr{P}}$ to denote the prefix $\{0,1\}^k 0\{*\}^{w-k-1}$, and $\overline{\mathscr{P}}$ to denote the prefix $\{0,1\}^k 1\{*\}^{w-k-1}$.

Given a one-dimensional packet classifier f on $\{*\}^w$, we use $f_{\mathscr{P}}$ to denote a packet classifier on $\mathscr{P}$ such that for any $x \in \mathscr{P}$, $f_{\mathscr{P}}(x) = f(x)$, and we use $f_{\mathscr{P}}^d$ to denote a similar packet classifier on $\mathscr{P}$ with the additional restriction that the final decision is d.

$C(f_{\mathscr{P}})$ denotes the minimum cost of a packet classifier t that is equivalent to $f_{\mathscr{P}}$, and $C(f_{\mathscr{P}}^d)$ denotes the minimum cost of a packet classifier t' that is equivalent to $f_{\mathscr{P}}$ and the decision of the last rule in t' is d.

Given a one-dimensional packet classifier f on $\{*\}^w$ and a prefix $\mathscr{P}$ where $\mathscr{P} \subseteq \{*\}^w$, f *is consistent on* $\mathscr{P}$ if and only if $\forall x,y \in \mathscr{P}$, $f(x) = f(y)$.

The dynamic programming solution to the weighted one-dimensional TCAM minimization problem is based on the following theorem. The proof of the theorem shows how to divide a problem into sub-problems and how to combine solutions to sub-problems into a solution to the original problem.

Theorem 2.1. *Given a one-dimensional packet classifier f on $\{*\}^w$, a prefix $\mathscr{P}$ where $\mathscr{P} \subseteq \{*\}^w$, the set of all possible decisions $\{d_1, d_2, \cdots, d_z\}$ where each decision d_i has a cost w_{d_i} $(1 \leq i \leq z)$, we have that*

$$C(f_{\mathscr{P}}) = \min_{i=1}^{z} C(f_{\mathscr{P}}^{d_i})$$

where each $C(f_{\mathscr{P}}^{d_i})$ is calculated as follows:
(1) If f is consistent on $\mathscr{P}$, then

$$C(f_{\mathscr{P}}^{d_i}) = \begin{cases} w_{f(x)} & \text{if } f(x) = d_i \\ w_{f(x)} + w_{d_i} & \text{if } f(x) \neq d_i \end{cases}$$

(2) If f is not consistent on $\mathscr{P}$, then

$$C(f_{\mathscr{P}}^{d_i}) = \min \begin{cases} C(f_{\underline{\mathscr{P}}}^{d_1}) + C(f_{\overline{\mathscr{P}}}^{d_1}) - w_{d_1} + w_{d_i}, \\ \cdots, \\ C(f_{\underline{\mathscr{P}}}^{d_{i-1}}) + C(f_{\overline{\mathscr{P}}}^{d_{i-1}}) - w_{d_{i-1}} + w_{d_i}, \\ C(f_{\underline{\mathscr{P}}}^{d_i}) + C(f_{\overline{\mathscr{P}}}^{d_i}) - w_{d_i}, \\ C(f_{\underline{\mathscr{P}}}^{d_{i+1}}) + C(f_{\overline{\mathscr{P}}}^{d_{i+1}}) - w_{d_{i+1}} + w_{d_i}, \\ \cdots, \\ C(f_{\underline{\mathscr{P}}}^{d_z}) + C(f_{\overline{\mathscr{P}}}^{d_z}) - w_{d_z} + w_{d_i} \end{cases}$$

Proof. (1) The base case is when f is consistent on $\mathscr{P}$. In this case, the minimum cost prefix packet classifier in $\{f_{\mathscr{P}}\}$ is clearly $\langle \mathscr{P} \to f(x) \rangle$, and the cost of this packet classifier is $w_{f(x)}$. Furthermore, for $d_i \neq f(x)$, the minimum cost prefix packet classifier in $\{f_{\mathscr{P}}\}$ with decision d_i in the last rule is $\langle \mathscr{P} \to f(x), \mathscr{P} \to d_i \rangle$ where the second rule is redundant. The cost of this packet classifier is $w_{f(x)} + w_{d_i}$.

(2) If f is not consistent on $\mathscr{P}$, divide $\mathscr{P}$ into $\underline{\mathscr{P}}$ and $\overline{\mathscr{P}}$. The crucial observation is that an optimal solution f^* to $\{f_{\mathscr{P}}\}$ is essentially an optimal solution f_1 to the sub-problem of minimizing $f_{\underline{\mathscr{P}}}$ appended with an optimal solution f_2 to the sub-problem of minimizing $f_{\overline{\mathscr{P}}}$. The only interaction that can occur between f_1 and f_2 is if their final rules have the same decision, in which case both final rules can be replaced with one final rule covering all of $\mathscr{P}$ with the same decision. Let d_x be the decision of the last rule in f_1 and d_y be the decision of the last rule in f_2. Then we can compose f^* whose last rule has decision d_i from f_1 and f_2 based on the following cases:

(A) $d_x = d_y = d_i$: In this case, f can be constructed by listing all the rules in f_1 except the last rule, followed by all the rules in f_2 except the last rule, and then the last rule $\mathscr{P} \to d_i$. Thus, $Cost(f) = Cost(f_1) + Cost(f_2) - w_{d_i}$.

(B) $d_x = d_y \neq d_i$: In this case, f can be constructed by listing all the rules in f_1 except the last rule, followed by all the rules in f_2 except the last rule, then rule $\mathscr{P} \to d_x$, and finally rule $\mathscr{P} \to d_i$. Thus, $Cost(f) = Cost(f_1) + Cost(f_2) - w_{d_x} + w_{d_i}$.

(C) $d_x \neq d_y, d_x = d_i, d_y \neq d_i$: We do not need to consider this case because $C(f_{\underline{\mathscr{P}}}^{d_i}) + C(f_{\overline{\mathscr{P}}}^{d_y}) = C(f_{\underline{\mathscr{P}}}^{d_i}) + (C(f_{\overline{\mathscr{P}}}^{d_y}) + w_{d_i}) - w_{d_i} \geq C(f_{\underline{\mathscr{P}}}^{d_i}) + C(f_{\overline{\mathscr{P}}}^{d_i}) - w_{d_i}$.

(D) $d_x \neq d_y, d_x \neq d_i, d_y = d_i$: Similarly, this case need not be considered.

(E) $d_x \neq d_y, d_x \neq d_i, d_y \neq d_i$: Similarly, this case need not be considered.

Figure 2.2 shows the illustration of a one-dimensional TCAM minimization problem, where the black bar denotes decision "accept" and the white bar denotes decision "discard". Figure 2.3 illustrates how the dynamic programming works on this example. Algorithm 4 shows the pseudocode for finding $C(f_{\mathscr{P}}^{d_i})$.

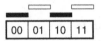

Fig. 2.2 An example one-dimensional TCAM minimization problem

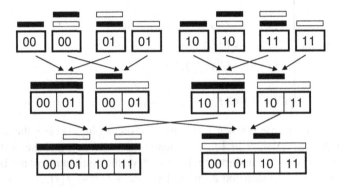

Fig. 2.3 Illustration of dynamic program

Input: A *Universe* they contains the color information of each prefix in the domain, a prefix
interval p, and dictionary of weights for each color.
Output: A dictionary that contains the cost of the optimal prefix solution for keys
$(Prefix, Color)$ where the prefix *Prefix* has the background color *Color*.

1 Let c be the colors at prefix p on *Universe* ;
2 Let *Colors* be the set of colors defined by *ColorWeights* ;
3 **if** *c is monochromatic* **then**
4   Let *answer* be an emtpy dictionary ;
5   **foreach** *color* $\in$ *Colors* **do**
6   **if** *color* $\neq c$ **then**
7             $answer[(p, color)] = ColorWeights[c] + ColorWeights[color]$;
8   **else**
9             $answer[(p, color)] = ColorWeights[c]$;
10   **end**
11   **end**
12 **else**
13   Split p into *low* and *high* ;
14   Let *answer* be a new dictionary that contains the keys from
ODPM(*Universe, low, ColorWeights*) and ODPM(*Universe, low, ColorWeights*) ;
15   **foreach** *color that is defined by ColorWeights* **do**
16         $answer[(p, color)] =$

$$\min_{cc \in Colors} \left\{ \begin{array}{l} answer[(lowPrefix, cc)] + answer[(highPrefix, cc)] \\ \qquad\qquad\qquad\qquad - ColorWeights[cc] \textbf{ if } color = cc \\ answer[(lowPrefix, cc)] + answer[(highPrefix, cc)] \\ \qquad\qquad - ColorWeights[cc] + ColorWeights[color] \textbf{ otherwise} \end{array} \right\}$$

        ;
17   **end**
18 **end**
19 **return** answer ;

Algorithm 4: ODPM(Universe, p, ColorWeights)

Chapter 3
Related Work

There is significant prior work on packet classification for both TCAM based packet classification and software based packet classification. While TCAM based systems are more immediately relevant, software based classification shares a degree of commonality with the sequential decomposition technique; however, differences in available hardware result in very different design decisions.

3.1 TCAM Based Classifiers

There is significant prior work on minimizing the TCAM space occupied by a *single classifier*. Such work falls into three broad categories: (1) *classifier minimization* (e.g.,, [Draves et al(1999)Draves, King, Venkatachary, and Zill, Suri et al(2003)Suri, Sandholm, and Warkhede, Applegate et al(2007)Applegate, Calinescu, Johnson, Karloff, Ligett, and Wang, Liu and Gouda(2005), Dong et al(2006)Dong, Banerjee, Wang, Agrawal, and Shukla, McGeer and Yalagandula(2009)]), which converts a given classifier to a semantically equivalent classifier that requires fewer TCAM entries; (2) *range encoding* (e.g.,, [Liu(2002), van Lunteren and Engbersen(2003), Pao et al(2006)Pao, Li, and Zhou, Lakshminarayanan et al(2005)Lakshminarayanan, Rangarajan, and Venkatachary, Bremler-Barr and Hendler(2007)]), which encodes the ranges (i.e.,, source port and destination port) in a manner that reduces range expansion; and (3) *circuit modification* (e.g.,, [Spitznagel et al(2003)Spitznagel, Taylor, and Turner]), which modifies TCAM circuits to accommodate range comparisons.

3.1.1 Classifier Minimization:

The basic idea is to convert a given packet classifier to another semantically equivalent packet classifier that requires fewer TCAM entries. Several classifier minimiza-

C. R. Meiners et al., *Hardware Based Packet Classification for High Speed Internet Routers*,   15
DOI 10.1007/978-1-4419-6700-8_3, © Springer Science+Business Media, LLC 2010

tion schemes have been proposed [Draves et al(1999)Draves, King, Venkatachary, and Zill,Suri et al(2003)Suri, Sandholm, and Warkhede,Applegate et al(2007)Applegate, Calinescu, Johnson, Karloff, Ligett, and Wang, Liu and Gouda(2005), Dong et al(2006)Dong, Banerjee, Wang, Agrawal, and Shukla]. The work in [Draves et al(1999)Draves, King, Venkatachary, and Zill, Suri et al(2003)Suri, Sandholm, and Warkhede,Applegate et al(2007)Applegate, Calinescu, Johnson, Karloff, Ligett, and Wang] focuses on one-dimensional and two dimensional packet classifiers.

Construction Optimal IP Tables

In [Draves et al(1999)Draves, King, Venkatachary, and Zill], Draves et al. present a polynomial algorithm for generating a minimum equivalent packet classifier for one-dimensional prefix match classifiers. Their algorithm, *Optimal Routing Table Constructor*(ORTC), works by reducing a longest matching prefix trie to its minimal representation via three traversals. This minimal trie can then be used to use to generate a minimum single field prefix classifier. A longest match prefix classifier can be trivially converted into a first match prefix classifier by sorting the rules such that the longer prefix rules appear before shorter prefix rules.

Compressing Two-Dimensional Routing Tables

In [Suri et al(2003)Suri, Sandholm, and Warkhede], Suri et al. present a polynomial time dynamic program that generates a minimum equivalent packet classifier for one-dimensional prefix classifiers. This dynamic program is equivalent to the dynamic program presented in Chapter 2.2. Furthermore, Suri et al. present a generalization of this dynamic program for two or more fields. They show that this generalization produces optimal two field classifiers when the the solution space of classifiers is restricted such that the predicates of any two rules in the a classifier are either disjoint, or one predicate is a subset of the other. However, these generalized algorithms have a significant time requirements, $O(N|DS|(w_1 \times \cdots \times w_d))$, where w_d is the number of bits used for F_i. As a result, the dynamic program ceases to be usable for more than two fields.

Complete Redundancy Detection in Firewalls

In [Liu and Gouda(2005)], Liu and Gouda, propose the first algorithm that is guaranteed to detect and remove a maximal set of redundant rules within a classifier. They propose two types of redundant rules, *upward redundant* rules and *downward redundant* rules. These two types of rules are shown to completely categorize the set of all redundant rules. Liu and Gouda's algorithm first uses an iterative FDD construction technique to remove all upward redundant rules, and it uses a different iterative FDD construction technique to remove all downward redundant rules. By removing both types of rules, the algorithm produces a classifier free of redundant rules. Note, that this algorithm is not guaranteed to remove the maximum number of redundant rules since there can be interdependencies between redundant rules. However, this algorithm is effective for all types of classifiers, and its efficiency scales well as the number of fields in a classifier increases.

Packet Classifiers in Ternary CAMs can be Smaller

In [Dong et al(2006)Dong, Banerjee, Wang, Agrawal, and Shukla], Dong et al. propose the first algorithm that modifies rules within a classifier in an attempt to reduce the effects of range expansion. They propose four types of operations: *trimming* rules, *expanding* rules, *merging* rules, and *adding* rules. The basic idea of their algorithm is that by trimming or expanding the space covered by a predicate, the range expansion for each rule can be reduced. They propose a two stage algorithm that first trims the predicate space of every rule and then expands the predicate space of each rule going from last to first. This algorithm is significant in that it accommodates classifiers with more than two fields. However, it is unknown whether or not the algorithm is optimal given a one-dimensional classifier. Furthermore, The algorithm requires repeated applications for a classifier to converge upon a minimal set of rules and depends heavily upon repeated applications of the redundancy removal technique found in [Liu and Gouda(2005)]. This suggests that their algorithm requires a significant amount of computational overhead.

Compressing Rectilinear Pictures and Minimizing Access Control Lists

In [Applegate et al(2007)Applegate, Calinescu, Johnson, Karloff, Ligett, and Wang], Applegate et al. propose an optimal solution for two field classifiers composed entirely of *strip rules*. Strip rules have a wild card for at least one field. However, while this work is of theoretical interest, it does not scale to d-dimensional classifiers, and it is not clear that packet classifiers can be efficiently represented with strip rules

Minimizing Rulesets for TCAM Implementation

In [McGeer and Yalagandula(2009)], McGeer and Yalgandula show that finding the minimal number of TCAM rules to represent a classifier is NP-hard. They show this property by reducing circuit equivalence to TCAM rule minimization. However, their reduction does not produce a feasible method for optimal TCAM rule minimization. The author, instead, propose a heuristic based approximation algorithm for TCAM classifiers with exactly two decisions.

3.1.2 Range Encoding:

The basic idea is to first encode ranges that appear in a classifier and store the encoded rules in a TCAM. When a packet comes, the packet needs to be preprocessed so that the resulting encoded packet can be used as a search key for the TCAM. Previous range encoding schemes fall into two categories: database independent encoding schemes [Lakshminarayanan et al(2005)Lakshminarayanan, Rangarajan, and Venkatachary, Bremler-Barr and Hendler(2007)], where the encoding of each rule is independent of other rules in the classifier, and database dependent encoding schemes [Liu(2002), van Lunteren and Engbersen(2003), Pao et al(2006)Pao, Li, and Zhou], where the encoding of each rule may depend on other rules in the classifier. The advantage of database independent encoding schemes is that they allow fast incremental updates to the classifier since each rule is encoded independently. However, database dependent schemes have the potential for better space savings since they can utilize the low number of unique ranges that appear in real life classifiers to achieve lower range expansion.

Algorithms for Advanced Packet Classification

Lakshminarayanan et al. propose a scheme called fence encoding, which encodes interval ranges as a range of unary numbers [Lakshminarayanan et al(2005)Lakshminarayanan, Rangarajan, and Venkatachary]. All ranges under fence encoding have an expansion factor of one, which implies that all ranges can be encoded with one rule, but the number of unary bits required for each rule is prohibitive since a field with length w requires 2^w bits per rule. To reduce the required number of bits in a rule, Lakshminarayan et al. proposed the technique called DIRPE, which compresses the size of fence encodings at the expense of increasing the average expansion ratio. DIRPE works by dividing a field into equally sized sub-fields, which are called *chunks*, and fence encoding these chunks. Selecting the number of chunks provides a trade-off between range expansion and TCAM entry width. The authors also propose a method of combining DIRPE with Liu's range encoding scheme found in [Liu(2002)] to handle ranges that have a large expansion factor under DIRPE. However, this combination negates DIRPE's ability to allow fast updates for classifiers.

Space-efficient TCAM-based Classification using Gray Coding

Bremler-Barr and Hendler, in [Bremler-Barr and Hendler(2007)], propose a scheme in which field domains are encoded using *binary reflected gray codes*(BRGC). While there is no advantage or disadvantage to using a BRGC for fields that contain only prefix ranges, using BRGC on non-prefix ranges breaks up the range in such a way that additional ternary bits can be used to eliminate some of the prefixes needed to represent a range. The result is that some of the prefixes ranges required to represent a range are merged together into a single ternary entry. The authors note that this encoding technique is especially effective for small ranges and name their encoding algorithm *short range gray encoding* or SGRE. Since SGRE does not require any additional TCAM bits to encode ranges, Bremler-Barr and Hendler also propose a method of combining SGRE with Liu's range encoding technique. Like DIRPE this combination negates SGRE's ability to support fast updates for classifiers, but it allows for the technique to concisely encode ranges with large expansion factors under SRGE.

Efficient Mapping of Range Classifiers into Ternary-CAM

In [Liu(2002)], Liu proposes an encoding method that designates specific ternary bits within each TCAM entry to represent a specific range. A packet field is encoded via an SRAM lookup table that maps each field value to a codeword that has a designated bit set to 1 if and only if the value is an element of the corresponding range. Each rule's range predicate can then be encoded such that the designated bit is set to 1 and every other bit is set to $*$. This technique eliminates range expansion completely; however, it also requires n bits per TCAM entry when a classifier has n unique ranges in a field. This technique quickly becomes impractical as the number of unique ranges within a field increases. To combat the explosive growth in required bits, Liu proposes splitting a field domain into k disjoint ranges such that each disjoint range intersects with a small number of unique ranges. Since $\lceil \log k + 1 \rceil$ bits are needed to encode these disjoint ranges, this scheme allows for a field to be encoded using $\lceil \log k + 1 \rceil + n'$ bits where n' is the maximum number of unique ranges that intersect with a given disjoint range. Using this scheme means that rule predicates that intersects with more than one disjoint range must be replicated for each intersection. To manage the trade off between rule expansion and bit expansion, Liu proposes a heuristic algorithm that repeatedly finds and merges the pair of disjoint ranges that reduces rule expansion the most. These merges continue until a budget of b bits is exhausted.

Fast and Scalable Packet Classification

In [van Lunteren and Engbersen(2003)], van Lunteren and Engbersen propose an encoding method similar to [Liu(2002)]. In their encoding scheme, they control the required number of bits by partitioning the unique ranges into l layers. The ranges within each layer are then broken into disjoint ranges so that each layer can be encoded in $\lceil \log n_i \rceil$ bits where n_i is the number of disjoint ranges in layer i. Each field then become the concatenated encoding for each layer. The authors also note that if a disjoint range r in one layer contains a disjoint range r' in another layer, this information can be used to reduce the number of bits needed to encode r''s layer. Unfortunately, no algorithms are given for partitioning unique ranges into layers.

An Encoding Scheme for TCAM-based Packet Classification

In [Pao et al(2006)Pao, Li, and Zhou], Pao et al. propose an encoding algorithm called *prefix inclusion encoding*(PIC). PIC utilizes van Lunteren and Engbersen's observation that containment information for one layer can reduce the number of bits required to encode ranges in the next layer. That is the scheme produces a series of l layers $L_1, \ldots, L_l$ such that each disjoint range in L_i is a subset of a single range of L_{i-1} for $i \in \{2, \ldots, l\}$. With this property, PIC can encode a field predicate into a compact prefix range. PIC was designed for encoding fields with prefix ranges and large domains such as the source IP for IPv6 headers. However, the authors do suggests techniques for adapting their scheme to encode range fields. These techniques require breaking overlapping ranges into disjoint ranges, which in turn requires that encoded rules are replicated in a manner similar to other encoding techniques.

3.1.3 Circuit Modification:

The basic idea is to modify TCAM circuits to accommodate range comparisons. For example, Spitznagel et al. proposed adding comparators at each entry level to better accommodate range matching [Spitznagel et al(2003)Spitznagel, Taylor, and Turner]. While this research direction is important, such solutions are hard to deploy due to high cost [Lakshminarayanan et al(2005)Lakshminarayanan, Rangarajan, and Venkatachary], and modified TCAMs may be less applicable to applications other than packet processing.

3.2 Software Based Techniques

The simplest software based technique for packet classification is a linear search, which has excellent storage requirement but becomes too slow for wire speed packet classification for even modest sized classifiers. As a results, a rich body of software based packet classification techniques have been developed [Singh et al(2003)Singh, Baboescu, Varghese, and Wang, Qiu et al(2001)Qiu, Varghese, and Suri, Taylor and Turner(2005), Lakshman and Stiliadis(1998), Gupta and McKeown(1999a), Feldmann and Muthukrishnan(2000), boe scu and Varghese(2001), Woo(2000), Baboescu et al(2003)Baboescu, Singh, and Varghese, Srinivasan et al(1998)Srinivasan, Varghese, Suri, and Waldvogel, iva san et al(1999)iva san, Suri, and Varghese], and an extensive survey of these techniques can be found in [Taylor(2005)]. These techniques trade storage space for an improvement in search time via special preprocessing of the classifier rules. Techniques can be partitioned into two categories: parallel decomposition and decision tree classification.

3.2.1 Parallel decomposition

The objective of parallel decomposition techniques [Lakshman and Stiliadis(1998), boe scu and Varghese(2001), Srinivasan et al(1998)Srinivasan, Varghese, Suri, and Waldvogel,iva san et al(1999)iva san, Suri, and Varghese,Gupta and McKeown(1999a), Taylor and Turner(2005)] is to break the classification process into several steps that can performed in parallel. The above techniques perform the decomposition along the field boundaries of a packet header. This in effect allows for fast and efficient single field classification solutions to encode each field in parallel. These new values are then composed via one or more additional classifications stages to yield a correct classification.

High-speed Policy-based Packet Forwarding using Efficient Multi-dimensional Range Matching

In [Lakshman and Stiliadis(1998)], Lakshman and Stiliadis propose encoding each field's value into a bitmap that specifies a containment relationship among values and rules [Lakshman and Stiliadis(1998)]. This bitmap indicates whether or not an encoded value intersects with a given rule's field predicate. Once each field is encoded, this method uses customized parallel AND gates to perform an intersection of these bitmaps and ultimately finds the first matching rule. This technique is effective; however it requires a bit line for each rule in the classifier and must be implemented on customized hardware.

Scalable Packet Classification

In [boe scu and Varghese(2001)], Baboescu and Varghese improve on the above technique by observing that for classifiers with a low occurrence of wildcards, bitmaps will be sparely populated with 1's. They group bits within each bitmap into chunks and represent each chuck with a single bit which is the logical OR of all the bits within the chunk. This allows the second stage to skip the comparison of a significant number of bits. For classifiers that have a low occurrence of wildcards, this technique is very effective at reducing the number of memory access needed to perform the second stage processing; however, this reduction is diminished once wildcards occur more frequently within a classifier.

Fast and Scalable Layer Four Switching

In [Srinivasan et al(1998)Srinivasan, Varghese, Suri, and Waldvogel], Srinivasan et al. propose an encoding method called *cross-producting* that assigns a unique number to each maximal disjoint range within a classifier field and constructs a lookup table for the cross product of the numbers associated with each field. This technique is fast; however, its storage requirements multiplicatively increases as the number of fields and ranges increases. As a result, the authors only intend crossproducting for small classifiers with two fields.

Packet Classification on Multiple Fields

In [Gupta and McKeown(1999a)], Gupta and McKeown propose an encoding method called Recursive Flow Classification (RFC) that is an optimized version of the cross-producting scheme. This uses recursive cross-producting tables to reduce the space requirements of regular cross-producting tables. Furthermore, they map disjoint ranges that are contained by the same set of rules into a single value. RFC's mapping tables define an equivalence relation; however, this equivalence relation is less general than the domain compression technique discussed in Chapter 8, so they are unable to achieve a maximum compression for each field domain in most cases. Furthermore, the recursive cross-producting scheme requires a significant amount of space to store in memory.

Packet Classification using Tuple Space Search

In [iva san et al(1999)iva san, Suri, and Varghese], Srinivasan et al. propose a tuple based search approach. This approach transforms each rule predicate with d fields into a d-tuple, which is in essence a hash of the predicate. The idea is that this initial hashing divides the search space into regions that can be searched in parallel. Perfect hashing functions are used to find exact matches in each tuple's search space. The authors propose two methods of determining appropriate tuples to search. The first method is an exhaustive search of each tuple, and the second uses a set pruning trie for each field that returns a set of candidate tuples. With set pruning, the intersections of the results from each field is the set of tuple spaces that need to be searched.

Scalable Packet Classification using Distributed Crossproducting of Field Labels

In [Taylor and Turner(2005)], Taylor and Turner propose the Distributed Crossproducting of Field Labels (DCFL) method that assigns each locally unique range within a field a locally unique number. Each field value is encoded into a set of numbers, which represents the ranges that contains the value. These sets are crossproducted together and then intersected with the set of unique tuples generated from the classifier's field predicates. The resulting intersection provides a list of rules that the packet header matches. The authors optimize this technique by incrementally performing the crossproduct and filtering the intermediate results after each incremental crossproduct. This optimization can dramatically reduce that number of false positive tuples that are generated. Since overlapping ranges diminish the incremental crossproducts' ability to keep the number of candidate matches low, the technique's performance depends on classifiers having a low number of overlapping ranges.

3.2.2 Decision Trees

Decision tree techniques [Gupta and McKeown(1999b), Singh et al(2003)Singh, Baboescu, Varghese, and Wang, Woo(2000), Qiu et al(2001)Qiu, Varghese, and Suri, Feldmann and Muthukrishnan(2000), Baboescu et al(2003)Baboescu, Singh, and Varghese] use tree structures to successively prune the search space to a single rule or a small number of rules, which are then searched linearly to find a match. Decision tree methods such as HiCuts [Gupta and McKeown(1999b)] and Hypercuts [Singh et al(2003)Singh, Baboescu, Varghese, and Wang] are similar in flavor to our sequential decomposition approach in that they use a sequence of searches where each search uses a portion of the packet predicate to classify a packet. However, software-based methods are constrained by a complex tradeoff among how many searches need to be performed, the time required to perform a search and the space required to store the data structure that facilitates the search. In the worst case, these methods require many searches, slow searches, or tremendous amounts of memory.

Classification Using Hierarchical Intelligent Cuttings

In [Gupta and McKeown(1999b)], Gupta and McKeown present a decision tree algorithm called *HiCuts*. This algorithm builds a decision tree similar to an unordered FDD with the following differences: Each node makes a decision based on a partition of a field's domain, and leaves are allowed to store a list of rules. The rationale for both of these decisions is derived directly from the limitations of SRAM lookup methods. The authors implement each node as a lookup table so that the next node in the tree can be found in constant time. However, since a field domain of size 2^{32} is prohibitively large, the field domain must be *cut* into subsets to limit the size of each tree node. This technique successively prunes the set of candidate rules with the decision tree until the set of rules is below a certain threshold. Once this threshold is reached, the list of remaining rules is stored in the leaf at the end of the decision path. The rationale for this decision is to save storage space since small lists usually result in big subtrees. The authors also present a parameterized construction algorithm that allows the user to trade maximum lookup time for storage space.

Packet Classification using Multidimensional Cutting

In [Singh et al(2003)Singh, Baboescu, Varghese, and Wang], Singh et al. present *HyperCuts*, which is an improvement upon HiCuts. The authors contribution is to allow each node in the decision tree to build a multidimensional lookup table from cuts in multiple fields. This improvement allows for a more effective pruning of the list of candidate rules. The authors show that HyperCuts significantly improves upon the performance of HiCuts.

A Modular Approach to Packet Classification

In [Woo(2000)], Woo uses a three stage approach to classifying packets. The first stage is a lookup table that distributes packet value among a set of decision trees by matching m bits within a rule predicate. These decision trees are binary trees where the nodes select the appropriate bit within the predicate to determine which edge to follow. The second stage traverses the decision tree until the third stage is reached when a leaf in the decision tree is found. These leaves contain a list of one or more candidate matches that is searched sequentially until a match is found. One key assumption in Woo's work is that each classifier predicate needs to be transformed into a ternary bit string. This assumption implies that classifiers with significant amounts of range expansion will degrade the storage efficiency of this technique.

Fast Firewall Implementations for Software-based and Hardware-based Routers

In [Qiu et al(2001)Qiu, Varghese, and Suri], Qiu et al. revisit two trie-based lookup schemes for packet classification that have been traditionally dismissed as being inefficient and show that for real packet classifiers, they offer predictable classification speeds. Longest prefix matching tries are an efficient data structure for performing an exact match for a single field packet; however, once packet classifiers requires multiple field packets, some packets will not match against the longest prefix in all dimensions. The first technique that they examine uses a backtracking search on a multi-field trie to find every candidate rule. They provide a set of optimizations for the multi-field trie that speeds up the backtracking search; however, the number of memory accesses required to classify each packet range from 117 to 196 for real-life classifiers. The second technique that they examine uses set pruning tries, which enumerate all decision paths so that each packet value can only follow a single path. The authors also propose two compression algorithms that help to reduce the storage requirements for set pruning tries and backtracking tries. Set pruning tries outperform backtracking search at the expense of additional memory storage requirements; however, experimental results suggest that backtracking tries offer a better performance for storage trade off. The experimental results suggest that these techniques offer a 2 to 5 times speedup over linear search.

Tradeoffs for Packet Classification

In [Feldmann and Muthukrishnan(2000)], Feldmann and Muthukrishnan propose building lookup-up trees similar to HiCuts; however, instead of using a lookup table at each node, they employ an inverted lookup tree call a *Fat Inverted Segment*(FIS) tree to store a complete set of cuts of each field. This technique allows for a more compact representation of the classifier, but it can significantly increasing the number memory accesses needed to classify a packet when compared to HiCuts or HyperCuts.

Packet Classification for Core Routers: Is there an Alternative to CAMs?

In [Baboescu et al(2003)Baboescu, Singh, and Varghese], Baboescu et al. propose the *Extended Grid-of-Tries*(EGT) technique. EGT uses a two-field trie to prune the candidate rule list and uses a path compression algorithm to minimize the amount of memory needed to store the trie. For core router tables, EGT provides reasonable performance; however, EGT's performance depends on the structural properties of the core routing tables. Packets classifiers used in other applications (e.g., firewalls) may not have acceptable performance with EGT.

Part I
Equivalent Transformation Techniques

Consider the following TCAM Minimization Problem: *given a packet classifier, how can we generate another semantically equivalent packet classifier that requires the least number of TCAM entries?* Two packet classifiers are (semantically) equivalent if and only if they have the same decision for every packet. For example, the two packets classifiers in Tables 1.1 and 1.2 are equivalent; however, the one in Table 1.1 requires 900 TCAM entries, and the one in Table 1.2 requires only 6 TCAM entries.

Solving this problem helps to address the limitations of TCAMs. As we reduce the number of TCAM entries required, we can use smaller TCAMs, which results in less board space and lower hardware cost. Furthermore, reducing the number of rules in a TCAM directly reduces power consumption and heat generation because the energy consumed by a TCAM grows linearly with the number of ternary rules it stores [Yu et al(2005)Yu, Lakshman, Motoyama, and Katz].

While the optimal solution to the above problem is conceivably NP-hard, in this thesis, we propose a practical algorithmic solution using two techniques. Our first technique, TCAM Razor, generates new but equivalent classifiers, whereas our second technique, all-match redundancy removal, finds a set of rules that can be safely removed from a classifier

Chapter 4
TCAM Razor

TCAM Razor consists of the following four basic steps. First, convert a given packet classifier to a reduced decision diagram, which is the canonical representation of the semantics of the given packet classifier. Second, for every nonterminal node in the decision diagram, minimize the number of prefixes associated with its outgoing edges using dynamic programming. Third, generate rules from the decision diagram. Last, remove redundant rules. As an example, running our algorithms on the packet classifier in Table 1.1 will yield the one in Table 1.2.

The solution is named "TCAM Razor" following the principle of Occam's razor: *"Of two equivalent theories or explanations, all other things being equal, the simpler one is to be preferred."* In our context, of all packet classifiers that are equivalent, the one with the least number of TCAM entries is preferred.

4.1 Multi-dimensional TCAM Minimization: The Basics

In this section, we present TCAM Razor, our algorithm for minimizing multi-dimensional prefix packet classifiers. A key idea behind TCAM Razor is processing one dimension at a time using the weighted one-dimensional TCAM minimization algorithm in Section 2.2 to greedily identify a local minimum for the current dimension. Although TCAM Razor is not guaranteed to achieve a global minimum across all dimensions, it does significantly reduce the number of prefix rules in real-life packet classifiers.

4.1.1 Conversion to Firewall Decision Diagrams

To facilitate processing a packet classifier one dimension at a time, we first convert a given packet classifier to an equivalent reduced *Firewall Decision Diagram* (FDD)

C. R. Meiners et al., *Hardware Based Packet Classification for High Speed Internet Routers,*   31
DOI 10.1007/978-1-4419-6700-8_4, © Springer Science+Business Media, LLC 2010

[Gouda and Liu(2007)]. Given a packet classifier f_1, we can construct an equivalent FDD f_2 using the FDD construction algorithm in [Liu and Gouda(2004)].

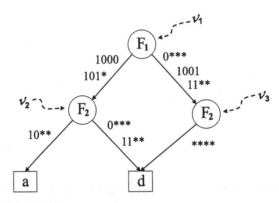

Fig. 4.1 A firewall decision diagram

4.1.2 Multi-dimensional TCAM Minimization

We start the discussion of our greedy solution by examining the reduced FDD in Figure 4.1. We first look at the subgraph rooted at node v_2. This subgraph can be seen as representing a one-dimension packet classifier over field F_2. We can use the weighted one-dimensional TCAM minimization algorithm in Section 2.2 to minimize the number of prefix rules for this one-dimensional packet classifier. The algorithm takes the following 3 prefixes as input:

$10**$ (with decision accept and cost 1),
$0***$ (with decision discard and cost 1),
$11**$ (with decision discard and cost 1).

The one-dimensional TCAM minimization algorithm will produce a minimum (one-dimensional) packet classifier of two rules as shown in Table 4.1.

Rule #	F_1	Decision
1	$10**$	accept
2	$****$	discard

Table 4.1 A minimum packet classifier corresponding to v_2 in Fig. 4.1

Similarly, from the subgraph rooted at node v_3, we can get a minimum packet classifier of one rule as shown in Table 4.2.

Rule #	F_1	Decision
1	****	discard

Table 4.2 A minimum packet classifier corresponding to v_3 in Fig. 4.1

Next, we look at the root v_1. As shown in Figure 4.2, we view the subgraph rooted at v_2 as a decision with a multiplication factor or cost of 2, and the subgraph rooted at v_3 as another decision with a cost of 1. Thus, the graph rooted at v_1 can be thought of as a "virtual" one-dimensional packet classifier over field F_1 where each child has a multiplicative cost.

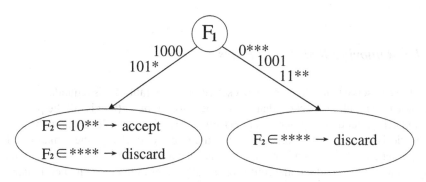

Fig. 4.2 "Virtual" one-dimensional packet classifier

Now we are ready to use the one-dimensional TCAM minimization algorithm in Section 2.2 to minimize the number of rules for this "virtual" one-dimensional packet classifier. The algorithm takes the following 5 prefixes and associated costs as input:

    1000   (with decision v_2 and cost 2),
    101∗   (with decision v_2 and cost 2),
    0∗∗∗   (with decision v_3 and cost 1),
    1001   (with decision v_3 and cost 1),
    11∗∗   (with decision v_3 and cost 1),

Running the weighted one-dimensional TCAM minimization algorithm on the above input will produce the "virtual" one-dimensional packet classifier of three rules as shown in Table 4.3.

Combining the "virtual" packet classifier in Table 4.3 and the two packet classifiers in Table 4.1 and 4.2, we get a packet classifier of 4 rules as shown in Table 4.4.

Rule #	F_1	Decision
1	1001	go to node v_3
2	10**	go to node v_2
3	****	go to node v_3

Table 4.3 A minimum packet classifier corresponding to v_1 in Fig. 4.1

Rule #	F_1	F_2	Decision
1	1001	****	discard
2	10**	10**	accept
3	10**	****	discard
4	****	****	discard

Table 4.4 Packet classifier generated from the FDD in Figure 4.1

4.1.3 Removing Redundant Rules

Next, we observe that rule r_3 in the packet classifier in Table 4.4 is redundant. If we remove rule r_3, all the packets that used to be resolved by r_3 (that is, all the packets that match r_3 but do not match r_1 and r_2) are now resolved by rule r_4, and r_4 has the same decision as r_3. Therefore, removing rule r_3 does not change the semantics of the packet classifier. Redundant rules in a packet classifier can be removed using the algorithms in [Liu and Gouda(2005)] or the algorithm in the next chapter. Finally, after removing redundant rules, we get a packet classifier of 3 rules from the FDD in Figure 4.1.

4.1.4 The Algorithm

To summarize, TCAM Razor, our multi-dimensional TCAM minimization algorithm, consists of the following four steps:

1. Convert the given packet classifier to an equivalent FDD.
2. Use the FDD reduction algorithm described in the next section to reduce the size of the FDD. This step will be explained in more detail in the next section.
3. Generate a packet classifier from the FDD in the following bottom up fashion. For every terminal node, assign a cost of 1. For a non-terminal node v with z outgoing edges $\{e_1, \cdots, e_z\}$, formulate a one-dimensional TCAM minimization problem as follows. For every prefix $\mathscr{P}$ in the label of edge e_j, $(1 \leq j \leq z)$, we set the decision of $\mathscr{P}$ to be j, and the cost of $\mathscr{P}$ to be the cost of the node that edge e_j points to. For node v, we use the weighted one-dimensional TCAM minimization algorithm in Section 2.2 to compute a one-dimensional prefix packet classifier with the minimum cost. We then assign this minimum cost to the cost of node v. After the root node is processed, generate a packet classifier using the prefixes

computed at each node in a depth first traversal of the FDD. The cost of the root indicates the total number of prefix rules in the resulting packet classifier.

4. Remove all the redundant rules from the resulting packet classifier.

4.1.5 TCAM Update

Packet classification rules periodically need to be updated. The common practice for updating rules is to run two TCAMs in tandem where one TCAM is used while the other is updated [Lekkas(2003)]. TCAM Razor is compatible with this current practice. Because TCAM Razor is efficient and the resultant TCAM lookup table is small, TCAM updating can be efficiently performed by rerunning TCAM Razor on the updated rules. When rules are frequently added to a classifier, we suggest the following lazy update strategy. First, after running TCAM Razor, store the resulting rules in the lower portion of the TCAM. Let n denote the total number of entries in the TCAM, m denote the total number of TCAM entries needed by a packet classifier after applying Razor, and let array T denote the TCAM. Initially, the m entries are stored from $T[n-m]$ to $T[n-1]$. When a new rule r needs to be added to the classifier, we first perform range expansion on r. Let m_1 be the number of prefix rules that are created. We store these rules in locations $T[n-m-m_1]$ to $T[n-m-1]$. As new rules are added, this process continues until the TCAM is filled up. Thus, TCAM Razor only needs to run periodically rather than when each new rule is added.

4.2 Multi-dimensional TCAM Minimization: The Optimization Techniques

In this section, we discuss the following two optimization techniques that we implemented to reduce the running time and memory usage of TCAM Razor: lazy copying in FDD construction and hashing in FDD reduction.

4.2.1 Lazy Copying in FDD Construction

The FDD construction algorithm in [Liu and Gouda(2004)] performs deep copying of subgraphs when splitting edges. This is inefficient in terms of both running time and memory usage. In TCAM Razor, we use the technique of lazy copying, which is explained as follows. Consider the subgraph (of an FDD) in Figure 4.3. The root of this subgraph is v, and v has k outgoing edges $e_1, e_2, \cdots, e_k$, which point to the subgraphs $g_1, g_2, \cdots, g_k$ respectively. When we need to make another copy of this subgraph, instead of making a deep copy of the whole subgraph, we only make another copy of the root of the subgraph. Let v' denote the new node. Node v' has the same label as v, and also has k outgoing edges $e'_1, e'_2, \cdots, e'_k$, where each e'_i has the same label $I(e_i)$ as e_i, and also points to the same subgraph g_i that e_i points to.

Each time we need to modify a node v, we first need to check its in-degree (i.e.,, the number of edges that point to v): if its indegree is 1, then we can directly modify v; if its indegree is greater than 1, then we need to first make a lazy copy of the subgraph rooted at v, and then modify the new node v'. To the outside, lazy copying looks like deep copying, but it reduces unnecessary copying of subgraphs (and promotes the sharing of common subgraphs) in the constructed FDD as much as possible.

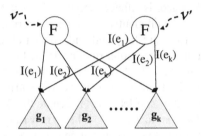

Fig. 4.3 Lazing copying of subgraphs

Fig. 4.4 shows the process of appending rule $(F_1 \in 0000) \wedge (F_2 \in 010*) \wedge (F_3 \in 0***) \rightarrow d$ to node v_1 of the partial FDD in the upper left side of the figure. A partial FDD is a diagram that has all the properties of an FDD except the completeness property.

In step (a), we split the single edge leaving v_1 into two edges, where v_5 is a shallow copy of v_2. In step (b), we further split the edge labeled 01** into two edges, where v_6 is a shallow copy of v_4. In step (c), we add the edge labeled 00** to v_6.

The pseudocode for the lazy copying based FDD construction algorithm is in Algorithm 5.

4.2.2 Hashing in FDD Reduction

To further reduce the number of rules generated by our algorithm, after we convert a packet classifier to an equivalent FDD, we need to reduce the size of the FDD. An FDD is *reduced* if and only if it satisfies the following three conditions: (1) no two nodes are isomorphic; (2) no two nodes have more than one edge between them; (3) no node has only one outgoing edge. Two nodes v and v' in an FDD are *isomorphic* if and only if v and v' satisfy one of the following two conditions: (1) both v and v' are terminal nodes with identical labels; (2) both v and v' are nonterminal nodes

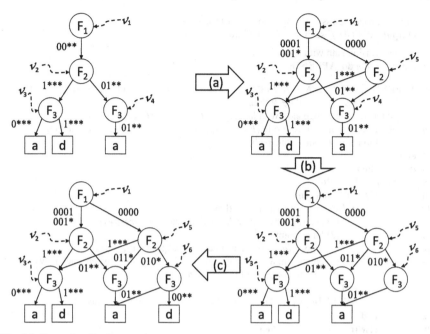

Fig. 4.4 Example of lazing copying

and there is a one-to-one correspondence between the outgoing edges of v and the outgoing edges of v' such that every pair of corresponding edges have identical labels and they both point to the same node.

We next show an example where FDD reduction helps to reduce the number of prefix rules generated from an FDD. Consider the two equivalent FDDs in where Figure 4.5a is non-reduced and Figure 4.5b is reduced. If we run our multi-dimensional TCAM minimization algorithm on the two FDDs, we will produce 4 prefix rules as shown in Table 4.5(a) and 2 prefix rules as shown in Table 4.5(b), respectively.

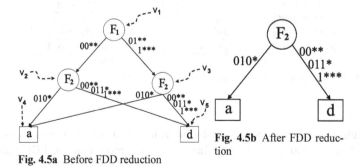

Fig. 4.5a Before FDD reduction

Fig. 4.5b After FDD reduction

Input: A packet classifier f of a sequence of rules $\langle r_1, \cdots, r_n \rangle$
Output: An FDD f' such that f and f' are equivalent

1 build a decision path with root v from rule r_1;
2 **for** $i := 2$ *to* n **do** **APPEND**(v, r_i);

3 **APPEND**$(v, (F_m \in S_m) \wedge \cdots \wedge (F_d \in S_d) \to \langle dec \rangle$) /*$F(v) = F_m$ and $E(v) = \{e_1, \cdots, e_k\}$*/
4 **if** $(S_m - (I(e_1) \cup \cdots \cup I(e_k))) \neq \emptyset$ **then**
5   add an outgoing edge e_{k+1} with label $S_m - (I(e_1) \cup \cdots \cup I(e_k))$ to v;
6   build a decision path from rule $(F_{m+1} \in S_{m+1}) \wedge \cdots \wedge (F_d \in S_d) \to \langle dec \rangle$, and make e_{k+1}
          point to the first node in this path;
7 **end**
8 **if** $m < d$ **then**
9   **for** $j := 1$ *to* k **do**
10   **if** $I(e_j) \subseteq S_m$ **then**
11   **if** *Indegree*$(Target(e_j)) > 1$ **then**
12   (1) create a new node v' labeled the same as $Target(e_j)$;
13   (2) let e_j point to v';
14   /*suppose $Target(e_j)$ has h outgoing edges $\varepsilon_1, \varepsilon_2, \cdots, \varepsilon_h$*/
15   (3) create h new outgoing edges $\varepsilon'_1, \varepsilon'_2, \cdots, \varepsilon'_h$ for v', where each new edge
                                 ε'_t point to $Target(\varepsilon_t)$ for $1 \leq t \leq h$;
16   **end**
17   **APPEND**$(Target(e_j), (F_{m+1} \in S_{m+1}) \wedge \cdots \wedge (F_d \in S_d) \to \langle dec \rangle)$;
18   **end**
19   **else if** $I(e_j) \cap S_m \neq \emptyset$ **then**
20   (1) create a new node v' labeled the same as $Target(e_j)$;
21   (2) create a new edge e from v to v' with label $I(e_j) \cap S_m$;
22   (3) replace the label of e_j by $I(e_j) - S_m$;
23   /*suppose $Target(e_j)$ has h outgoing edges $\varepsilon_1, \varepsilon_2, \cdots, \varepsilon_h$*/
24   (4) create h new outgoing edges $\varepsilon'_1, \varepsilon'_2, \cdots, \varepsilon'_h$ for v', where each new edge ε'_t
                         point to $Target(\varepsilon_t)$ for $1 \leq t \leq h$;
25   (5) **APPEND**$(Target(e_j), (F_{m+1} \in S_{m+1}) \wedge \cdots \wedge (F_d \in S_d) \to \langle dec \rangle$);
26   **end**
27   **end**
28 **end**

Algorithm 5: Lazy Copying Based FDD Construction

#	F_1	F_2	Decision
1	00**	010*	accept
2	00**	****	discard
3	****	010*	accept
4	****	****	discard

Rules generated from Figure 4.5a

#	F_1	F_2	Decision
1	****	010*	accept
2	****	****	discard

Rules generated from Figure 4.5b

(a)   (b)

Table 4.5 Rules from the FDDs in Figure 4.5a and Figure 4.5b

A brute force deep comparison algorithm for FDD reduction was proposed in [Gouda and Liu(2007)]. In TCAM Razor, we use a more efficient FDD reduction algorithm that processes the nodes level by level from the terminal nodes to the root node using signatures to speed up comparisons. This algorithm works as follows.

Starting from the bottom level, at each level, we compute a signature for each node at that level. For a terminal node v, set v's signature to be its label. For a non-terminal node v, suppose v has k children $v_1, v_2, \cdots, v_k$, in increasing order of signature ($Sig(v_i) < Sig(v_{i+1})$ for $1 \leq i \leq k-1$), and the edge between v and its child v_i is labeled with E_i, a sequence of non-overlapping prefixes in increasing order. Set the signature of node v as $Sig(v) = h(Sig(v_1), E_1, \cdots, Sig(v_k), E_k)$ where h is a one-way and collision resistant hash function such as MD5 [Rivest(1992)] and SHA-1 [Eastlake and Jones(2001)]. For any such hash function h, given two different input x and y, the probability of $h(x) = h(y)$ is extremely small.

After we have assigned signatures to all nodes at a given level, we search for isomorphic subgraphs as follows. For every pair of nodes v_i and v_j ($1 \leq i \neq j \leq k$) at this level, if $Sig(v_i) \neq Sig(v_j)$, then we can conclude that v_i and v_j are not isomorphic; otherwise, we explicitly determine if v_i and v_j are isomorphic. If v_i and v_j are isomorphic, we delete node v_j and its outgoing edges, and redirect all the edges that point to v_j to point to v_i. Further, we eliminate double edges between node v_i and its parents. For example, the signatures of the non-root nodes in Figure 4.5a are computed as follows:

$Sig(v_4) = a$
$Sig(v_5) = d$
$Sig(v_2) = h(Sig(v_4), 010*, Sig(v_5), 00**, 011*, 1***)$
$Sig(v_3) = h(Sig(v_4), 010*, Sig(v_5), 00**, 011*, 1***)$

At the end, for any nonterminal node v, if v has only one outgoing edge, we remove v and redirect the incoming edge of v to v's single child. As this step does not affect the number of rules generated from the FDD, we can skip it in practice.

Chapter 5
Bit Weaving

Prior TCAM-based classifier compression schemes [Draves et al(1999)Draves, King, Venkatachary, and Zill, Suri et al(2003)Suri, Sandholm, and Warkhede, Applegate et al(2007)Applegate, Calinescu, Johnson, Karloff, Ligett, and Wang, Dong et al(2006)Dong, Banerjee, Wang, Agrawal, and Shukla, Liu and Gouda(to appear), Meiners et al(2007)Meiners, Liu, and Torng] suffer from one fundamental limitation: they only produce prefix classifiers, which means they all miss some opportunities for compression. A prefix classifier is a classifier in which every rule is a prefix rule. In a prefix rule, each field is specified as a prefix bit string (e.g.,, 01**) where *s all appear at the end. In a ternary rule, each field is a ternary bit string (e.g.,, 0**1) where * can appear at any position. Every prefix rule is a ternary rule, but not vice versa. Because all previous compression schemes can only produce prefix rules, they miss the compression opportunities created by non-prefix ternary rules.

Bit weaving is a new TCAM-based classifier compression scheme that is not limited to producing prefix classifiers. The basic idea of bit weaving is simple: adjacent TCAM entries that have the same decision and have a hamming distance of one (i.e.,, differ by only one bit) can be merged into one entry by replacing the bit in question with *. Bit weaving applies two new techniques, *bit swapping* and *bit merging*, to first identify and then merge such rules together. Bit swapping first cuts a rule list into a series of partitions. Within each partition, a single permutation is applied to each rule's predicate to produce a reordered rule predicate, which forms a single prefix where all *'s are at the end of the rule predicate. This single prefix format allows us to use existing dynamic programming techniques [Meiners et al(2007)Meiners, Liu, and Torng, Suri et al(2003)Suri, Sandholm, and Warkhede] to find a minimal TCAM table for each partition in polynomial time. Bit merging then finds and merges mergeable rules from each partition. After bit merging, we revert all ternary strings back to their original bit permutation to produce the final TCAM table. We name our solution bit weaving because it manipulates bit ordering in a ternary string much like a weaver manipulates the position of threads.

The example in Figure 5.1 shows that bit weaving can further compress a minimal prefix classifier. The input classifier has 5 prefix rules with three decisions (0,

C. R. Meiners et al., *Hardware Based Packet Classification for High Speed Internet Routers*, 41
DOI 10.1007/978-1-4419-6700-8_5, © Springer Science+Business Media, LLC 2010

1, and 2) over two fields F_1 and F_2, where each field has two bits. Bit weaving compresses this minimal prefix classifier with 5 rules down to 3 ternary rules as follows.
First, it cuts the input prefix classifier into two partitions which are the first two rules
and the last three rules, respectively. Second, it swaps bit columns in each partition
to make the two-dimensional rules into one-dimension prefix rules. In this example,
in the second partition, the second and the fourth columns are swapped. We call the
above two steps *bit swapping*. Third, we treat each partition as a one-dimensional
prefix rule list and generate a minimal prefix representation. In this example, the
second partition is minimized to 2 prefix rules. Fourth, in each partition, we detect
and merge rules that differ by a single bit. In the first partition, the two rules are
merged. We call this step *bit merging*. Finally, we revert each partition back to its
original bit order. In this example, for the second partition after minimization, we
swap the second and the fourth columns again to recover the original bit order. The
final output is a ternary packet classifier with only 3 rules.

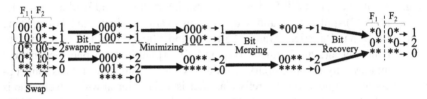

Fig. 5.1 Example of the bit weaving approach

To implement bit weaving, we must solve several challenging technical problems. First, we need to develop an algorithm that partitions a rule list into the least
number of partitions. Second, we must develop an algorithm that permutes the bit
columns within each partition to produce one-dimensional prefix rule lists. Third,
we must adapt the existing one-dimensional prefix rule list minimization algorithms
to minimize *incomplete* one-dimensional rule lists. Finally, we must develop algorithms to detect and then merge mergeable rules within each partition.

Our bit weaving approach has many significant benefits. First, it is the first
TCAM compression method that can create non-prefix classifiers. All previous
compression methods [Liu and Gouda(2005), Liu et al(2008)Liu, Meiners, and
Zhou, Meiners et al(2007)Meiners, Liu, and Torng, Dong et al(2006)Dong, Banerjee, Wang, Agrawal, and Shukla] generate only prefix classifiers. This restriction to
prefix format may miss important compression opportunities. Second, it is the first
efficient compression method with a polynomial worst-case running time with respect to the number of fields in each rule. Third, it is orthogonal to other techniques,
which means that it can be run as a pre/post-processing routine in combination with
other compression techniques. In particular, bit weaving complements TCAM Razor [Meiners et al(2007)Meiners, Liu, and Torng] nicely. In our experiments on
real-world classifiers, bit weaving outperforms TCAM Razor on classifiers that do
not have significant range expansion. Fourth, it supports fast incremental updates to
classifiers.

The rest of this chapter proceeds as follows. We define bit swapping in Section 5.1 and bit merging in Section 5.2. In Section 5.3, we discuss how bit weaving supports incremental updates, how bit weaving can be composed with other compression methods, and the complexity bounds of bit weaving.

5.1 Bit Swapping

In this section, we present a new technique called *bit swapping*. It is the first part of our bit weaving approach.

5.1.1 Prefix Bit Swapping Algorithm

Definition 5.1 (Bit-swap). A *bit-swap* β of a length m ternary string t is a permutation of the m ternary bits; that is, β rearranges the order of the ternary bits of t. The resulting permuted ternary string is denoted $\beta(t)$. $\square$

For example, if β is permutation 312 and string t is 0*1, then $\beta(t) = 10*$. For any length m string, there are $m!$ different bit-swaps. Bit-swap β is a *prefix bit-swap* of t if the permuted string $\beta(t)$ is in prefix format. Let $P(t)$ denote the set of prefix bit-swaps for t: specifically, the bit-swaps that move the $*$ bits of t to the end of the string.

A bit-swap β can be applied to a list ℓ of ternary strings $\langle t_1, \ldots, t_n \rangle$ where ℓ is typically a list of consecutive rules in a packet classifier. The resulting list of permuted strings is denoted as $\beta(\ell)$. Bit-swap β is a prefix bit-swap for ℓ if β is a prefix bit-swap for every string t_i in list ℓ for $1 \leq i \leq n$. Let $P(\ell)$ denote the set of prefix bit-swaps for list ℓ. It follows that $P(\ell) = \cap_{i=1}^{n} P(t_i)$.

Prefix bit-swaps are useful for two main reasons. First, we can minimize prefix rule lists using algorithms in [Meiners et al(2007)Meiners, Liu, and Torng, Suri et al(2003)Suri, Sandholm, and Warkhede, Draves et al(1999)Draves, King, Venkatachary, and Zill]. Second, prefix format facilitates the second key idea of bit weaving, bit merging (Section 5.2). After bit merging, the classifier is reverted to its original bit order, which typically results in a non-prefix format classifier.

Unfortunately, many lists of string ℓ have no prefix bit-swaps which means that $P(\ell) = \emptyset$. For example, the list $\langle 0*, *0 \rangle$ does not have a prefix bit-swap. We now give the necessary and sufficient conditions for $P(\ell) \neq \emptyset$ after defining the following notation.

Given that each ternary string denotes a set of binary strings, we define two new operators for ternary strings: $\hat{0}(x)$ and $\sqsubseteq$. For any ternary string x, $\hat{0}(x)$ denotes the resulting ternary string where every 1 in x is replaced by 0. For example, $\hat{0}(1*)=0*$. For any two ternary strings x and y, $x \sqsubseteq y$ if and only if $\hat{0}(x) \subseteq \hat{0}(y)$. For example, $1* \sqsubseteq 0*$ because $\hat{0}(1*)=0*=\{00, 01\} \subseteq \{00, 01\}=\hat{0}(0*)$.

Definition 5.2 (Cross Pattern). Given two ternary strings t_1 and t_2, a cross pattern on t_1 and t_2 exists if and only if $(t_1 \not\sqsubseteq t_2) \wedge (t_2 \not\sqsubseteq t_1)$. In such cases, we say that t_1 *crosses* t_2. □

We first observe that bit swaps have no effect on whether or not two strings cross each other.

Observation 5.1. *Given two ternary strings, t_1 and t_2, and a bit-swap β, $t_1 \sqsubseteq t_2$ if and only if $\beta(t_1) \sqsubseteq \beta(t_2)$, and $t_1 \sqsubseteq t_2$ if and only if $\beta(t_1) \sqsubseteq \beta(t_2)$.* □

Theorem 5.1. *Given a list $\ell = \langle t_1, \ldots, t_n \rangle$ of n ternary strings, $P(\ell) \neq \emptyset$ if and only if no two ternary strings t_i and t_j $(1 \leq i < j \leq n)$ cross each other.* □

Proof. *(implication)* It is given that there exists a prefix bit-swap $\beta \in P(\ell)$. Suppose that string t_i crosses string t_j. According to Observation 5.1, $\beta(t_i)$ crosses $\beta(t_j)$. This implies that one of the two ternary strings $\beta(t_i)$ and $\beta(t_j)$ has a $*$ before a 0 or 1 and is not in prefix format. Thus, β is not in $P(\ell)$, which is a contradiction.

(converse) It is given that no two ternary strings cross each other. It follows that we can impose a total order on the ternary strings in ℓ using the relation $\sqsubseteq$. Note, there may be more than one total order if $t_i \sqsubseteq t_j$ and $t_j \sqsubseteq t_i$ for some values of i and j. Let us reorder the ternary strings in ℓ according to this total order; that is, $t'_1 \sqsubseteq t'_2 \sqsubseteq \cdots \sqsubseteq t'_{n-1} \sqsubseteq t'_n$. Any bit swap that puts the $*$ bit positions of t'_1 last, preceded by the $*$ bit positions of $t'_2, \ldots$, preceded by the $*$ bit positions of t'_n, finally preceded by all the remaining bit positions will be a prefix bit-swap for ℓ. Thus, the result follows.

Theorem 5.1 gives us a simple algorithm for detecting whether a prefix bit-swap exists for a list of ternary strings. If a prefix bit-swap exists, the proof of Theorem 5.1 gives us an *algorithm for constructing a prefix bit-swap* as shown in Algorithm 6. The algorithm sorts bit columns in an increasing order by the number of strings that have a $*$ in that column.

Before we formally present our bit swapping algorithm, we define the concepts of *bit matrix* and *decision array* for a possibly incomplete rule list (i.e.,, there may exist a packet that none of the n rules matches). Any list of n rules defines a bit matrix $M[1..n, 1..b]$ and a decision array $D[1..n]$, where for any $1 \leq i \leq n$ and $1 \leq j \leq b$, $M[i, j]$ is the j-th bit in the predicate of the i-th rule and $D[i]$ is the decision of the i-th rule. Conversely, a bit matrix $M[1..n, 1..b]$ and a decision array $D[1..n]$ also uniquely defines a rule list. Given a bit matrix $M[1..n, 1..b]$ and a decision array $D[1..n]$ defined by a rule list, our bit swapping algorithm swaps the columns in M such that for any two columns i and j in the resulting bit matrix M' where $i < j$, the number of *s in the i-th column is less than or equal to the number of *s in the j-th column. Figure 5.2(a) shows a bit matrix and Figure 5.2(b) shows the resulting bit matrix after bit swapping. Let L_1 denote the rule list defined by M and D, and let L_2 denote the rule list defined by M' and D. Usually, L_1 will not be equivalent to L_2. This is not an issue. The key is that if we revert the bit-swap on any rule list L_3 that is equivalent to L_2, the resulting rule list L_4 will be equivalent to L_1.

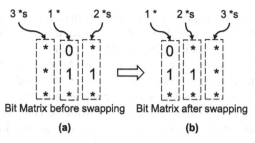

Fig. 5.2 Example of bit-swapping

Input: A classifier C of n rules $\langle r_1,\ldots,r_n\rangle$ where each rule has b bits.
Output: A classifier C' that is C after a valid prefix bit-swap.

1 Let $M[1\ldots n, 1\ldots b]$ and $D[1\ldots n]$ be the bit matrix and decision array of C;
2 Let $B = \langle (i,j) | 1 \le i \le b$ where j is the number of *'s in $M[1\ldots n, i]\rangle$;
3 Sort B in ascending order of each pair's second value;
4 Let M' be a copy of M;
5 **for** $k := 1$ to b **do**
6   Let $(i,j) = B[k]$;
7      $M'[1\ldots n, k] := M[1\ldots n, i]$;
8 **end**
9 Output C' defined by M' and D;

Algorithm 6: Finds a prefix bit-swap

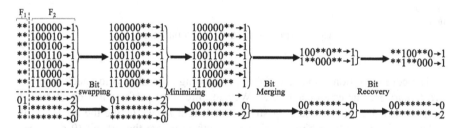

Fig. 5.3 Applying bit weaving algorithm to an example classifier

5.1.2 Minimal Cross-Free Classifier Partitioning Algorithm

Given a classifier $\mathbb{C}$, if $P(\mathbb{C}) = \emptyset$, we cut $\mathbb{C}$ into partitions where each partition has no cross patterns and thus has a prefix bit-swap. We treat classifier $\mathbb{C}$ as a list of ternary strings by ignoring the decision of each rule.

Given an n-rule classifier $\mathbb{C} = \langle r_1,\ldots,r_n\rangle$, a *partition* $\mathbb{P}$ on $\mathbb{C}$ is a list of consecutive rules $\langle r_i,\ldots,r_j\rangle$ in $\mathbb{C}$ for some i and j such that $1 \le i \le j \le n$. A *partitioning*, $\mathbb{P}_1,\ldots,\mathbb{P}_k$, of $\mathbb{C}$ is a series of k partitions on $\mathbb{C}$ such that the concatenation of $\mathbb{P}_1,\ldots,\mathbb{P}_k$ is $\mathbb{C}$. A partitioning is *cross-free* if and only if each partition has no cross patterns. Given a classifier $\mathbb{C}$, a cross-free partitioning with k partitions is *minimal* if and only if any partitioning of $\mathbb{C}$ with $k-1$ partitions is not cross-free. The goal of classifier partitioning is to find a minimal cross-free partitioning for a given classifier. We then apply independent prefix bit-swaps to each partition.

We give an algorithm, depicted in Algorithm 7, that finds a minimal cross-free partitioning for a given classifier. At any time, we have one active partition. The initial active partition is the last rule of the classifier. We consider each rule in the classifier in reverse order and attempt to add it to the active partition. If the current rule crosses any rule in the active partition, that partition is completed, and the active partition is reset to contain only the new rule. We process rules in reverse order to facilitate efficient incremental update (Section 5.3.2). New rules are more likely to be added to the front of a classifier than at the end. It is not hard to prove that this algorithm produces a minimal cross-free partitioning for any given classifier.

Input: A list of n rules $\langle r_1, \ldots, r_n \rangle$ where each rule has b bits.
Output: A list of partitions.

1 Let P be the current partition (empty list), and L be a list of partitions (empty);
2 **for** $i := n$ **to** 1 **do**
3   **if** r_i *introduces a cross pattern in P* **then**
4   Append P to the head of L;
5         $P := \langle r_i \rangle$;
6   **else**
7   Append r_i to the head of P;
8   **end**
9 **end**
10 return L;

Algorithm 7: Find a minimal partition

The core operation in our cross-free partitioning algorithm is to check whether two ternary strings cross each other. We can efficiently perform this check based on Theorem 5.2. For any ternary string t of length m, we define the *bit mask* of t, denoted $M(t)$, to be a binary string of length m where the i-th bit ($0 \leq i < m$) $M(t)[i] = 0$ if $t[i] = *$ and $M(t)[i] = 1$ otherwise. For any two binary strings a and b, we use $a \&\& b$ to denote the resulting binary string of the bitwise logical AND of a and b.

Theorem 5.2. *For any two ternary string t_1 and t_2, t_1 does crosses t_2 if and only if $M(t_1) \&\& M(t_2)$ is different from both $M(t_1)$ and $M(t_2)$.* $\square$

For example, given two ternary strings $t_1 = 01{*}0$ and $t_2 = 101{*}$, whose bit masks are $M(t_1) = 1101$ $M(t_1) = 1110$, we have $M(t_1) \&\& M(t_2) = 1100$. Therefore, $t_1 = 01{*}0$ crosses $t_2 = 101{*}$ because $M(t_1) \&\& M(t_2) \neq M(t_1)$ and $M(t_1) \&\& M(t_2) \neq M(t_2)$.

Figure 5.3 shows the execution of our bit weaving algorithm on an example classifier. Here we describe the bit swapping portion of that execution. The input classifier has 10 prefix rules with three decisions (0, 1, and 2) over two fields F_1 and F_2, where F_1 has two bits, and F_2 has six bits. We begin by constructing a maximal cross-free partitioning of the classifier by starting at the last rule and working up-ward. We find that the seventh rule introduces a cross pattern with the eighth rule according to Theorem 5.2. This results in splitting the classifier into two partitions. Second, we perform bit swapping on each partition, which converts each partition into a list of one-dimensional prefix rules.

5.1.3 Partial List Minimization Algorithm

We now describe how to minimize each bit-swapped partition where we view each partition as a list of 1-dimensional prefix rules. If a list of 1-dimensional prefix rules is complete (i.e.,, any packet has a matching rule in the list), we can use the algo-rithms in [Suri et al(2003)Suri, Sandholm, and Warkhede, Draves et al(1999)Draves, King, Venkatachary, and Zill] to produce an equivalent minimal prefix rule list. However, the rule list in a partition is often incomplete; that is, there exist pack-ets that do not match any rule in the partition.

Instead, we adapt the Weighted 1-Dimensional Prefix List Minimization Algo-rithm in [Meiners et al(2007)Meiners, Liu, and Torng] to minimize a partial 1-dimensional prefix rule list L over field F as follows. Let $\{d_1, d_2, \cdots, d_z\}$ be the set of all the decisions of the rules in L. Create a default rule r^* that matches all packets and assign it decision d_{z+1}. Append r^* to L to create a new classifier L'. As-sign each decision in $\{d_1, d_2, \cdots, d_z\}$ a weight of 1 and the decision d_{z+1} a weight of $|D(F)|$, the size of the domain F. Finally, run the weighted 1-dimensional prefix list minimization algorithm in [Meiners et al(2007)Meiners, Liu, and Torng] on L'. Given our weight assignment for decisions, we know the final rule of the resulting classifier L'' will still be r^* and that this rule will be the only one with decision d_{z+1}. Remove this final rule r^* from L'', and the resulting prefix list is the minimal partial prefix list that is equivalent to L.

Continuing the example from Figure 5.3, we use the partial prefix list minimiza-tion algorithm to minimize each partition to its minimal prefix representation. In this example, this step eliminates one rule from the bottom partition.

5.2 Bit Merging

In this section, we present bit merging, the second part of our bit weaving approach. T he fundamental idea behind bit merging is to repeatedly find in a classifier two ternary strings that differ only in one bit and replace them with a single ternary string where the differing bit is $*$.

5.2.1 Definitions

Two ternary strings t_1 and t_2 are *ternary adjacent* if they differ only in one bit, i.e.,, their hamming distance [Hamming(1950)] is one. The ternary string produced by replacing the one differing bit by a $*$ in t_1 (or t_2) is called the *ternary cover* of t_1 and t_2. For example, $0**$ is the ternary cover of $00*$ and $01*$. We call the process of replacing two ternary adjacent strings by their cover *bit merging* or just merging. For example, we can merge $00*$ and $01*$ to form their cover $0**$.

We now define how to bit merge (or just merge) two rules. For any rule r, we use $\mathbb{P}(r)$ to denote the predicate of r. Two rules r_i and r_j are ternary adjacent if their predicates $\mathbb{P}(r_i)$ and $\mathbb{P}(r_j)$ are ternary adjacent. The merger of ternary adjacent rules r_i and r_j is a rule whose predicate is the ternary cover of $\mathbb{P}(r_i)$ and $\mathbb{P}(r_j)$ and whose decision is the decision of rule r_i. We give a necessary and sufficient condition where bit merging two rules does not change the semantics of a classifier.

Theorem 5.3. *Two rules in a classifier can be merged into one rule without changing the classifier semantics if and only if they satisfy the following three conditions: (1) they can be moved to be positionally adjacent without changing the semantics of the classifier; (2) they are ternary adjacent; (3) they have the same decision.* □

The basic idea of bit merging is to repeatedly find two rules in the same bit-swapped partition that can be merged based on the three conditions in Theorem 5.3. We do not consider merging rules from different bit-swapped partitions because any two bits from the same column in the two bit-swapped rules may correspond to different columns in the original rules.

5.2.2 Bit Merging Algorithm (BMA)

5.2.2.1 Prefix Chunking

To address the first condition in Theorem 5.3, we need to quickly determine what rules in a bit-swapped partition can be moved together without changing the semantics of the partition (or classifier). For any 1-dimensional minimum prefix classifier $\mathbb{C}$, let $\mathbb{C}^s$ denote the prefix classifier formed by sorting all the rules in $\mathbb{C}$ in decreasing order of prefix length. We prove that $\mathbb{C} \equiv \mathbb{C}^s$ if $\mathbb{C}$ is a 1-dimensional minimum prefix classifier in Theorem 5.4.

Before we introduce and prove Theorem 5.4, we first present Lemma 5.1. A rule r is *upward redundant* if and only if there are no packets whose first matching rule is r [Liu and Gouda(2005)]. Clearly, upward redundant rules can be removed from a classifier with no change in semantics.

Lemma 5.1. *For any two rules r_i and r_j $(i < j)$ in a prefix classifier $\langle r_1, \cdots, r_n \rangle$ that has no upward redundant rules, $\mathbb{P}(r_i) \cap \mathbb{P}(r_j) \neq \emptyset$ if and only if $\mathbb{P}(r_i) \subset \mathbb{P}(r_j)$.* □

Theorem 5.4. *For any one-dimensional minimum prefix packet classifier* $\mathbb{C}$, *we have* $\mathbb{C} \equiv \mathbb{C}^s$.

Proof. Consider any two rules r_i, r_j $(i < j)$ in $\mathbb{C}$. If the prefixes of r_i and r_j do not overlap (i.e., $\mathbb{P}(r_i) \cap \mathbb{P}(r_j) = \emptyset$), changing the relative order between r_i and r_j does not change the semantics of $\mathbb{C}$. If the prefixes of r_i and r_j do overlap (i.e., $\mathbb{P}(r_i) \cap \mathbb{P}(r_j) \neq \emptyset$), then according to Lemma 5.1, we have $\mathbb{P}(r_i) \subset \mathbb{P}(r_j)$. This means that $\mathbb{P}(r_i)$ is strictly longer than $\mathbb{P}(r_j)$. This implies that r_i is also listed before r_j in $\mathbb{C}^s$. Thus, the result follows. $\square$

Based on Theorem 5.4, given a minimum sized prefix bit-swapped partition, we first sort the rules in decreasing order of their prefix length. Second, we further partition the rules into *prefix chunks* based on their prefix length. By Theorem 5.4, the order of the rules within each prefix chunk is irrelevant.

5.2.2.2 Bit-Mask Grouping

To address the second condition in Theorem 5.3, we need to quickly determine what rules are ternary adjacent. Based on Theorem 5.5, we can significantly reduce our search space by searching for mergeable rules only among the rules which have the same bit mask and decision.

Theorem 5.5. *Given a list of rules such that the rules have the same decision and no rule's predicate is a proper subset of another rule's predicate, if two rules are mergeable, then the bit masks of their predicates are the same.*

Proof. Suppose in such a list there are two rules r_i and r_j that are mergeable and have different bit masks. Because they are mergeable, $\mathbb{P}(r_i)$ and $\mathbb{P}(r_j)$ differ in only one bit. Because the bit masks are different, one predicate, say $\mathbb{P}(r_i)$, must have a $*$ and the other predicate, $\mathbb{P}(r_j)$, must have a 0 or 1 in that bit column. Thus, $\mathbb{P}(r_j) \subset \mathbb{P}(r_i)$, which is a contradiction.

5.2.2.3 Algorithm and Optimality

The bit merging algorithm (BMA) works as follows. BMA takes as input a minimum, possibly incomplete prefix classifier $\mathbb{C}$ that corresponds to a cross-free partition generated by bit swapping. BMA first creates classifier $\mathbb{C}^s$ by sorting the rules of $\mathbb{C}$ in decreasing order of their prefix length and partitions $\mathbb{C}^s$ into prefix chunks. Second, for each prefix chunk, BMA groups all the rules with the same bit mask and decision together, eliminates duplicate rules, and searches within each group for mergeable rules. The second step repeats until no group contains rules that can be merged. Let $\mathbb{C}'$ denote the output of the algorithm.

Figure 5.4 demonstrates how BMA works. On the leftmost side is the first partition from Figure 5.3. On the first pass, eight ternary rules are generated from the original seven. For example, the top two rules produce the rule $1000*0** \rightarrow 1$. These eight rules are grouped into four groups with identical bit masks. On the second pass, two unique rules are produced by merging rules from the four groups. Since each rule is in a separate group, no further merges are possible and the algorithm finishes. Algorithm 8 shows the general algorithm for BMA.

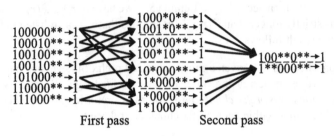

First pass   Second pass

Fig. 5.4 Example of Bit Merging Algorithm Execution

Input: A list I of n rules $\langle r_1, \ldots, r_n \rangle$ where each rule has b bits.
Output: A list of m rules.

1 Let S be the set of rules in I;
2 Let C be the partition of S such that each partition contains a maximal set of rules in S where each rule has an identical bitmask and decision;
3 Let OS be an empty set;
4 **for each** $c = \{r'_1, \ldots, r'_m\} \in C$ **do**
5   **for** $i := 1$ *to* $m - 1$ **do**
6   **for** $j := i + 1$ *to* m **do**
7   **if** $\mathbb{P}(r'_i)$ and $\mathbb{P}(r'_j)$ are ternary adjacent **then** Add the ternary cover of $\mathbb{P}(r'_i)$ and $\mathbb{P}(r'_j)$ to OS;
8   **end**
9   **end**
10 **end**
11 Let O be OS sorted in decreasing order of prefix length;
12 **if** $S = O$ **then**
13   return O;
14 **else**
15   return the result of BMA with O as input;
16 **end**

Algorithm 8: Bit Merging Algorithm

The correctness of this algorithm, $\mathbb{C}' \equiv \mathbb{C}$, is guaranteed because we only combine mergeable rules. We now prove that BMA is locally optimal as stated in Theorem 5.6.

Lemma 5.2. *During each execution of the second step, BMA never introduces two rules r_i and r_j such that $\mathbb{P}(r_i) \subset \mathbb{P}(r_j)$ where both r_i and r_j have the same decision.*
□

Lemma 5.3. *Consider any prefix chunk in* $\mathbb{C}^s$. *Let k be the length of the prefix of this prefix chunk. Consider any rule r in* $\mathbb{C}'$ *that was formed from this prefix chunk. The kth bit of r must be 0 or 1, not* $*$. □

Theorem 5.6. *The output of BMA,* $\mathbb{C}'$, *contains no pair of mergeable rules.*

Proof. Within each prefix chunk, after applying BMA, there are no pairs of mergeable rules for two reasons. First, by Theorem 5.5 and Lemma 5.2, in each run of the second step of the algorithm, all mergeable rules are merged. Second, repeatedly applying the second step of the algorithm guarantees that there are no mergeable rules in the end.

We now prove that any two rules from different prefix chunks cannot be merged. Let r_i and r_j be two rules from two different prefix chunks in $\mathbb{C}'$ with the same decision. Suppose r_i is from the prefix chunk of length k_i and r_j is from the prefix chunk of length k_j where $k_i > k_j$. By Lemma 5.3, the k_i-th bit of r_i's predicate must be 0 or 1. Because $k_i > k_j$, the k_i-th bit of r_j's predicate must be $*$. Thus, if r_i and r_j are mergeable, then r_i and r_j should only differ in the k_i-th bit of their predicates, which means $\mathbb{P}(r_i) \subset \mathbb{P}(r_j)$. This conflicts with Lemma 5.2. □

Continuing the example in Figure 5.3, we perform bit merging on both partitions to reduce the first partition to two rules. Finally, we revert each partition back to its original bit order. After reverting each partition's bit order, we recover the complete classifier by appending the partitions together. In Figure 5.3, the final classifier has four rules.

5.3 Discussion

5.3.1 Redundancy Removal

Our bit weaving algorithm uses the redundancy removal procedure in Chapter 6 as both the preprocessing and postprocessing step. We apply redundancy removal at the beginning because redundant rules may introduce more cross patterns. We apply redundancy removal at the end because our incomplete 1-dimensional prefix list minimization algorithm may introduce redundant rules across different partitions.

5.3.2 Incremental Classifier Updates

Classifier rules periodically need to be updated when networking services change. When classifiers are updated manually by network administrators, timing is not a concern and rerunning the fast bit weaving algorithm will suffice. When classifiers are updated automatically in an incremental fashion, fast updates may be very important.

Bit weaving supports efficient incremental classifier updates by confining change impact to one cross-free partition. An incremental classifier change is typically inserting a new rule, deleting an existing rule, or modifying a rule. Given a change, we first locate the cross-free partition where the change occurs by consulting a precomputed list of all the rules in each partition. Then we rerun the bit weaving algorithm on the affected partition. We may need to further divide the partition into two cross-free partitions if the change introduces a cross pattern. Note that deleting a rule never introduces cross patterns. We generated our partitions by processing rules in reverse order because new rules are most likely to be placed at the front of a classifier.

5.3.3 Composability of Bit Weaving

Bit weaving, like redundancy removal, never returns a classifier that is larger than its input. Thus, bit weaving, like redundancy removal, can be composed with other classifier minimization schemes. Since bit weaving is an efficient algorithm, we can apply it as a postprocessing step with little performance penalty. As bit weaving uses techniques that are significantly different than other compression techniques, it can often provide additional compression.

We can also enhance other compression techniques by using bit weaving, in particular bit merging, within them. Specifically, multiple techniques [Meiners et al(2007)Meiners, Liu, and Torng, Meiners et al(2008a)Meiners, Liu, and Torng, Meiners et al(2008c)Meiners, Liu, and Torng, Pao et al(2007)Pao, Zhou, Liu, and Zhang, Che et al(2008)Che, Wang, Zheng, and Liu] rely on generating single field TCAM tables. These approaches generate minimal prefix tables, but minimal prefix tables can be further compressed by applying bit merging. Therefore, every such technique can be enhanced with bit merging (or more generally bit weaving).

For example, TCAM Razor [Meiners et al(2007)Meiners, Liu, and Torng] compresses multiple field classifiers by converting a classifier into multiple single field classifiers, finding the minimal prefix classifiers for these classifiers, and then constructing a new prefix field classifier from the prefix lists. A natural enhancement is to use bit merging to convert the minimal prefix rule lists into smaller non-prefix rule lists. In our experiments, bit weaving enhanced TCAM Razor yields significantly better compression results than TCAM Razor alone.

Range encoding techniques [Liu(2002), van Lunteren and Engbersen(2003), Pao et al(2007)Pao, Zhou, Liu, and Zhang, Che et al(2008)Che, Wang, Zheng, and Liu,Meiners et al(2008c)Meiners, Liu, and Torng] can also be enhanced by bit merging. Range encoding techniques require lookup tables to encode fields of incoming packets. When such tables are stored in TCAM, they are stored as single field classifiers. Bit merging offers a low cost method to further compress these lookup tables. Our results show that bit merging significantly compresses the lookup tables formed by the topological transformation technique [Meiners et al(2008c)Meiners, Liu, and Torng].

5.3.4 Prefix Shadowing

During bit swapping, we produce a list of partial classifiers when we run the minimal cross-free partitioning algorithm. We then compress each partial classifier using the Weighted 1-Dimensional Prefix List Minimization Algorithm. During this step, when compressing a partial classifier, we take into account later partial classifiers with a high cost default rule. Specifically, we add a default rule to the partial classifier and give it high weight which forces the minimization algorithm to keep this rule as the last rule which can then be removed to produce a minimal partial classifier. We now show how we can further compress these partial classifiers by considering *prior* partial classifiers. We call this technique prefix shadowing.

Partition 1
$*000 \rightarrow d \Rightarrow *000 \rightarrow d$
$0111 \rightarrow d \Rightarrow 0111 \rightarrow d$
Partition 2
$010* \rightarrow a$
$0110 \rightarrow a \Rightarrow 0*** \rightarrow a$
$00** \rightarrow a$

Table 5.1 The second partition benefits from the prefix shadow of the first partition

Table 5.1 contains two partitions or partial classifiers that are both minimal according to the partial list minimization algorithm. Partition 1's bit swap is changed to partition 2's bit swap in order to directly compare the predicates of each partition's rules. We observe that if the rule $0111 \rightarrow d$ was present in partition 2, partition 2's three rules could be written as a single rule $0*** \rightarrow a$. Therefore, utilizing prefix coverage information from prior partitions can decrease the size of later partitions in the classifier.

We modify the partial list minimization algorithm to account for prefixes that are covered by prior partitions as follows. Let L be the list of rules in the partial classifier with original decisions $d_1, \ldots, d_z$. Let r^* represent the default rule with single decision d_{z+1} that accounts for all later partial classifiers. The original decisions $d_1, \ldots, d_z$ are assigned weight 1 whereas the decision d_{z+1} is assigned weight $|D(F)|$. Let $\hat{L}$ represent the list of all prefixes that are covered by prior classifiers, and let d_{z+2} be the single decision for all prefixes in $\hat{L}$. We give decision d_{z+2} weight 0 because all rules with decision d_{z+2} are free. Specifically, prefixes with decision d_{z+2} are covered by prior partial classifiers which makes them redundant. Thus, all rules with decision d_{z+2} can be safely removed from the given partial classifier without changing the semantics of the entire classifier. We create an input instance for the Weighted 1-Dimensional Prefix list Minimization Algorithm by concatenating $\hat{L}$ with L followed by r^*. For our example, this algorithm produces the single entry partial classifier in partition 2 instead of the three entry classifier.

5.4 Complexity Analysis of Bit Weaving

There are two computationally expensive stages to bit weaving: finding the minimal cross-free partition and bit merging. For analyzing both stages, let w be the number of bits within a rule predicate, and let n be the number of rules in the input. We show that bit merging's worst case time complexity is $O(w \times n^{2 \lg 3})$ (where $\lg 3 = \log_2 3$), which makes bit weaving the first polynomial-time algorithm with a worst-case time complexity that is independent of the number of fields in the input classifier.

Finding a minimal cross-free partition is composed of a loop of n steps. Each of these steps checks whether or not the adding the next rule to the current partition will introduce a cross pattern; this requires a linear scan comparing the next rule to each rule of the current partition. Each comparison takes $\Theta(w)$ time. In the worst case, we get scanning behavior similar to insertion sort which requires $\Theta(wn^2)$ time and $\Theta(wn)$ space.

We now analyze the complexity of bit merging, the most expensive stage of bit weaving. The key to our analysis is determining how many ternary covers are generated from our input of n rules. This analysis is complicated by the fact that the generation of ternary covers proceeds in rounds. We sidestep this complication by observing that if the total number of ternary covers generated in all rounds is $f(n)$, then the total space required by bit merging is $O(wf(n))$ and the total time required by bit merging is $O(w(f(n))^2)$. The time complexity follows because in the worst case, each ternary cover is compared against every other ternary cover to see if a new ternary cover can be created. This is an overestimate since we only compare the ternary covers generated in each round to each other, and we only compare ternary covers within the same prefix chunk to each other. We now show that the total number of ternary covers generated is $O(n^{\lg 3})$ which means that bit merging has a worst case time complexity of $O(wn^{2 \lg 3})$ and a worst case space complexity of $O(wn^{\lg 3})$.

$$0***0 \Rightarrow \begin{array}{l} 00000 \\ 00010 \\ 00100 \\ 00110 \\ 01000 \\ 01010 \\ 01100 \\ 01110 \end{array}$$

Table 5.2 The output of bit merging determines the input.

Based on Lemma 5.3, we restrict our attention to individual prefix chunks (where all the rules have the same decision) since we never merge rules from different prefix chunks. Furthermore, based again on Lemma 5.3, we assume that all input rules end with a 0 or 1 by eliminating all the $*$'s at the right end of all rules in this prefix chunk. We now perform our counting analysis by starting with the output of bit merging for this prefix chunk rather than the input to bit merging.

First consider the case where we have a single output rule with b *'s such as $0***0$. We observe that the number of input rules in this prefix chunk with the same decision must be exactly 2^b because the initial input is generated by the partial list minimization algorithm which only generates classifiers with prefix rules. Thus, the original input rules could not include non-prefix ternary rules like $000*0$. We illustrate this observation in Table 5.2 where we list all 8 input rules that must exist to produce the single output of $0***0$.

Pass			
0	1	2	3
	000*0		
00000	001*0	00**0	
00010	010*0	01**0	
00100	011*0		
00110	00*00		
	00*10	0*0*0	0***0
	01*00	0*1*0	
01000	01*10		
01010	0*000		
01100	0*010	0**00	
01110	0*100	0**10	
	0*110		
$\binom{3}{0}\times2^{3-0}$	$\binom{3}{1}\times2^{3-1}$	$\binom{3}{2}\times2^{3-2}$	$\binom{3}{3}\times2^{3-3}$
Ternary Covers			

Table 5.3 The set of all ternary covers generated per round

We now count the total number of ternary covers that are generated during each round of bit merging. If we consider the input set of rules as round 0, we observe that the number of ternary covers at round k is exactly $\binom{b}{k}\times2^{b-k}$ for $0\le k\le b$. Table 5.3 illustrates this property for the bit merging output $0***0$. If we sum over each round, the total number of ternary covers is exactly $\sum_{k=0}^{b}\binom{b}{k}2^{b-k}=3^b$ by the binomial theorem. Since the input size $n=2^b$, the total number of ternary covers is $3^b=(2^b)^{\lg3}=n^{\lg3}$.

We now consider the case where the output for a single prefix chunk (with the same decision) is q rules $R=\{r_i\mid1\le i\le q\}$ where $q>1$. Let b_i for $1\le i\le q$ be the number of *'s in output rule r_i. Let I_i be the set of input rules associated with rule r_i and let C_i be the set of ternary covers generated by I_i during the course of bit merging. For any subset $R'\subseteq R$, $I(R')=\cap_{r_i\in R'}I_i$ and $C(R')=\cap_{r_i\in R'}C_i$ where the intersection operator for a single set returns the same set. Stated another way, $I(R')$ is the set of input rules common to all the output rules in R', and $C(R')$ is the set of ternary covers common to all the output rules in R'. Let $c(R')$ be the ternary cover in $C(R')$ with the most stars. This ternary cover is uniquely defined for any R'. In particular, $c(R')$ would be the output rule if the input rules were exactly $I(R')$, and $C(R')$ would be exactly the set of ternary covers generated from $I(R')$. Let $b_{R'}$ be the number of stars in $c(R')$. This implies $I(R')$ contains exactly $2^{b_{R'}}$ input rules and $C(R')$ contains exactly $3^{b_{R'}}$ ternary covers.

Let $TI(R)$ be the total number of unique input rules associated with any output rule in R and let $TC(R)$ be the total number of unique ternary covers associated with any output rule in R. We now show how to compute $TI(R)$ and $TC(R)$ by applying the inclusion-exclusion principle. Let R^k for $1 \leq k \leq q$ be the set of subsets of R containing exactly k rules from R.

$$TI(R) = \sum_{k=1}^{w} (-1)^{k-1} \sum_{R' \in R^k} |I(R')| = \sum_{k=1}^{w} (-1)^{k-1} \sum_{R' \in R^k} 2^{b_{R'}}$$

$$TC(R) = \sum_{k=1}^{w} (-1)^{k-1} \sum_{R' \in R^k} |C(R')| = \sum_{k=1}^{w} (-1)^{k-1} \sum_{R' \in R^k} 3^{b_{R'}}$$

It follows that the total number of ternary covers generated is $O(n^{\lg 3})$.

For example, suppose $q = 2$ and $r_1 = 0{*}{*}{*}0$ and $r_2 = {*}{*}{*}00$. Then $c(R) = 0{*}{*}00$, $I(R) = \{00000, 00100, 01000, 01100\}$, and the remaining four elements of $C(R')$ are $\{00{*}00, 0{*}000, 01{*}00, 0{*}100\}$, and $TI(R) = 2^3 + 2^3 - 2^2 = 8 + 8 - 4 = 12$ and $TC(R) = 3^3 + 3^3 - 3^2 = 27 + 27 - 9 = 45$.

Chapter 6
All-Match Redundancy Removal

We present an all-match based complete redundancy removal algorithm. This is the first algorithm that attempts to solve first-match problems from an all-match perspective. We formally prove that the resulting packet classifiers have no redundant rules after running our redundancy removal algorithm.

We have improved upon [Liu and Gouda(2005)] in two ways. First, the redundancy theorem becomes simpler. The redundancy theorem in [Liu and Gouda(2005)] distinguishes upward and downward redundant rules, and detects them separately. In contrast, the redundancy theorem presented here gives a single criterion that can detect both upward and downward redundant rules. Second, the new redundancy removal algorithm is more efficient. The algorithm in [Liu and Gouda(2005)] scans a packet classifier twice and build FDDs twice in order to remove the two types of redundant rules. In comparison, the new algorithm only scans a packet classifier once and builds one all-match FDD with a cost similar cost to building an FDD. The new algorithm is about twice as efficient as the algorithm in [Liu and Gouda(2005)].

6.1 All-Match Based Redundancy Theorem

In this section, we introduce the concept of all-match FDDs and the all-match based redundancy theorem.

6.1.1 All-Match FDDs

Definition 6.1 (All-Match FDD). An *all-match FDD* t for a packet classifier f : $\langle r_1, r_2, \cdots, r_n \rangle$ over fields $F_1, \cdots, F_d$ is an FDD that has the following five properties:

1. Each node v is labeled with a packet field denoted $F(v)$. If v is a nonterminal node, then $F(v)$ is a packet field. If v is a terminal node, then $F(v)$ is a list of integer values $\langle i_1, i_2, \cdots, i_k \rangle$ where $1 \leq i_1 < i_2 \cdots < i_k \leq n$.

C. R. Meiners et al., *Hardware Based Packet Classification for High Speed Internet Routers*, 57
DOI 10.1007/978-1-4419-6700-8_6, © Springer Science+Business Media, LLC 2010

2. Each edge $e{:}u \rightarrow v$ is labeled with a nonempty set of integers, denoted $I(e)$, where $I(e)$ is a subset of the domain of u's label (i.e.,, $I(e) \subseteq D(F(u))$).
3. The set of all outgoing edges of a node v in t, denoted $E(v)$, satisfies the following two conditions:

   a. *Consistency*: $I(e) \cap I(e') = \emptyset$ for any two distinct edges e and e' in $E(v)$.
   b. *Completeness*: $\bigcup_{e \in E(v)} I(e) = D(F(v))$.

4. A directed path from the root to a terminal node is called a *decision path*. No two nodes on a decision path have the same label. Given a decision path $\mathscr{P}$: $(v_1 e_1 v_2 e_2 \cdots v_m e_m v_{m+1})$, the matching set of $\mathscr{P}$ is defined as the set of all packets that satisfy $(F(v_1) \in I(e_1)) \wedge (F(v_2) \in I(e_2)) \wedge \cdots \wedge (F(v_m) \in I(e_m))$. We use $M(\mathscr{P})$ to denote the matching set of $\mathscr{P}$.
5. For any decision path $\mathscr{P}$: $(v_1 e_1 v_2 e_2 \cdots v_m e_m v_{m+1})$ where $F(v_{m+1}) = \langle i_1, i_2, ..., i_k \rangle$ and for any rule $r_j (1 \le j \le n)$, if $M(\mathscr{P}) \cap M(r_j) \ne \phi$, then $M(\mathscr{P}) \subseteq M(r_j)$ and $j \in \{i_1, i_2, \cdots, i_k\}$. $\qquad \square$

Fig 6.2 shows an all-match FDD for the simple packet classifier in Fig 6.1. In this example, we assume every packet has only two fields F_1 and F_2, and the domain of each field is $[1, 10]$.

$$r_1 : F_1 \in [1, \;5] \wedge F_2 \in [1, 10] \rightarrow accept$$
$$r_2 : F_1 \in [1, \;5] \wedge F_2 \in [5, 10] \rightarrow accept$$
$$r_3 : F_1 \in [6, \;10] \wedge F_2 \in [1,3] \rightarrow discard$$
$$r_4 : F_1 \in [1, \;10] \wedge F_2 \in [1,10] \rightarrow discard$$

Fig. 6.1 A simple packet classifier

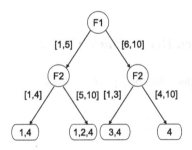

Fig. 6.2 An all-match FDD for the packet classifier in Fig 6.1

In an all-match FDD for a packet classifier f, for any decision path $\mathscr{P}$: $(v_1 e_1 v_2 e_2 \cdots v_m e_m v_{m+1})$ where $F(v_{m+1}) = \langle i_1, i_2, ..., i_k \rangle$, if a packet p satisfies this path $\mathscr{P}$, then $\{r_{i_1}, r_{i_2}, ..., r_{i_k}\}$ are exactly *all the rules in f that p matches*. This is why we call such a FDD an "all-match FDD".

6.1.2 The All-Match Based Redundancy Theorem

Before we present the All-Match Based Redundancy Theorem, we first prove the following lemma.

Lemma 6.1. *Let t be an all-match FDD for packet classifier $f : \langle r_1, r_2, \cdots, r_n \rangle$. For any rule r_i in f, let $\mathscr{P}_1, \mathscr{P}_2, \cdots, \mathscr{P}_h$ be all the decision paths whose terminal node contains r_i, then the following condition holds: $M(r_i) = \bigcup_{j=1}^{h} M(\mathscr{P}_j)$.* □

Proof:

(1) According to property 5 in the definition of all-match FDDs, we have $M(\mathscr{P}_j) \subseteq M(r_i)$ for every j ($1 \leq j \leq h$). Thus, we have $\bigcup_{j=1}^{h} M(\mathscr{P}_j) \subseteq M(r_i)$.

(2) Consider a packet p in $M(r_i)$. According to the consistency and completeness properties of all-match FDDs, there exists one and only one decision path that p matches. Let $\mathscr{P}$ be this decision path. Thus, we have $p \in M(r_i) \cap M(\mathscr{P})$. According to property 5 in the definition of all-match FDDs, i is in the label of $\mathscr{P}$'s terminal node. Thus, we have $\mathscr{P} \in \{\mathscr{P}_1, \mathscr{P}_2, \cdots, \mathscr{P}_h\}$. Therefore, we have $p \in \bigcup_{j=1}^{h} M(\mathscr{P}_j)$. Thus we get $M(r_i) \subseteq \bigcup_{j=1}^{h} M(\mathscr{P}_j)$. □

Theorem 6.1 (All-Match Based Redundancy Theorem). *Let t be an all-match FDD for packet classifier $f : \langle r_1, r_2, \cdots, r_n \rangle$. Rule r_i is redundant in f if and only if in all terminal nodes of t that have i as their first value, i is immediately followed by another integer j such that r_i and r_j have the same decision.*

Proof. (1) Suppose in all terminal nodes of t that have i as their first value, i is immediately followed by another integer j such that r_i and r_j have the same decision. We next prove that r_i is redundant in f.

We observe that removing a rule r_i only possibly affects the decisions for the packets in $M(r_i)$. Let $\mathscr{P}_1, \mathscr{P}_2, \cdots, \mathscr{P}_h$ be all the decision paths in t whose terminal node contains i. According to Lemma 6.1, we have $M(r_i) = \bigcup_{j=1}^{h} M(\mathscr{P}_j)$. Consider an arbitrary packet p in $M(r_i)$. Suppose we have $p \in M(\mathscr{P}_j)$. Let f' be the resulting packet classifier after removing r_i from f. To prove that r_i is redundant in f, we only need to prove $f(p) = f'(p)$. Let the label of the terminal node of $\mathscr{P}_j$ be $\langle i_1, i_2, \cdots, i_k \rangle$. Because $i \in \{i_1, i_2, \cdots, i_k\}$, there are two cases:

1. $i_1 \neq i$. In this case, r_{i_1} is the first rule in f that p matches. Thus, removing r_i does not affect the decision for p. In this case, we have $f(p) = f'(p)$.
2. $i_1 = i$, and r_i has the same decision as r_{i_2}. In f, r_i is the first rule that p matches. In f', r_{i_2} is the first rule that p matches. Because r_i and r_{i_2} has the same decision, we have $f(p) = f'(p)$ in this case.

Therefore, r_i is redundant in f.

(2) Suppose rule r_i is redundant in f and there exists a terminal node in t whose first two values are i followed by j, and r_i and r_j have different decisions. Let $\mathscr{P}$ denote the decision path from the root to this terminal node. Consider a packet $p \in M(\mathscr{P})$. Thus, r_i is the first rule that p matches in f and r_j is the first rule that p matches in f'. Because r_i and r_j have different decisions, we have $f(p) \neq f'(p)$.

This conflicts with the assumption that r_i is redundant. Therefore, if r_i is redundant in f, then in all terminal nodes of t that have i as their first value, i is immediately followed by another integer j such that r_i and r_j have the same decision.

6.2 All-Match Based Redundancy Removal

In this section, we first present an algorithm for constructing all-match FDDs from packet classifiers. Second, we present a redundancy removal algorithm based on Theorem 6.1. Third, we prove that the resulting packet classifier does not have any redundant rules.

6.2.1 The All-Match FDD Construction Algorithm

According to Theorem 6.1, in order to detect and remove redundant rules in a packet classifier, we first need to construct an all-match FDD for that packet classifier. The pseudocode for the all-match FDD construction algorithm is shown in Algorithms 9 and 10.

Input: A packet classifier $f : \langle r_1, r_2, \cdots, r_n \rangle$
Output: A all-match FDD t for packet classifier f

1 Build a path from rule r_1. Let v denote the root. The label of the terminal node is $\langle 1 \rangle$. ;
2 **for** $i = \{2, \ldots, n\} \in C$ **do**
3   APPEND($v, r_i, 1, i$) ;
4 **end**

Algorithm 9: All-Match FDD Construction Algorithm

Consider the packet classifier in Figure 6.1. The process of constructing the corresponding all-match FDD is shown in Figure 6.3.

6.2.2 The All-Match Based Redundancy Removal Algorithm

We first introduce two auxiliary lists that are used in the all-match based redundancy removal algorithm: containment list and residency list. Given an all-match FDD that has m terminal nodes, we assign a unique sequence number in $[1, m]$ to each terminal node. In the containment list, each entry consists of a terminal node sequence number and the rule sequence numbers contained in the terminal node. In the residency list, each entry consists of a rule sequence number and the set of terminal nodes

Input: A vertex v, a rule $(F_1 \in S_1) \wedge \cdots \wedge (F_d \in S_d) \rightarrow \langle dec \rangle$, a depth m, and a rule number i.
Output: v includes the rule $(F_1 \in S_1) \wedge \cdots \wedge (F_d \in S_d) \rightarrow \langle dec \rangle$ in its all-match structure.

/* $F(v) = F_m$ and $E(v) = \{e_1, \cdots, e_k\}$ */

1   **if** $m = d + 1$ **then**
2   Add i to the end of v's label. ;
3   **return**
4   **end**
5   **else if** $(S_m - (I(e_1) \cup \cdots \cup I(e_k))) \neq \emptyset$ **then**
6   Add an outgoing edge e_{k+1} with label $S_m - (I(e_1) \cup \cdots \cup I(e_k))$ to v ;
7   Build a decision path from $(F_{m+1} \in S_{m+1}) \wedge \cdots \wedge (F_d \in S_d) \rightarrow \langle dec \rangle$, and make e_{k+1} point to the first node in this path ;
8   Add i to the end of the label of the terminal node of this decision path ;
9   **end**
10 **for** $j := 1$ *to* k **do**
11   **if** $I(e_j) \subseteq S_m$ **then**
12   APPEND(e_j's target,$(F_1 \in S_1) \wedge \cdots \wedge (F_d \in S_d) \rightarrow \langle dec \rangle$,m+1,i);
13   **end**
14   Add one outgoing edge e to v, and label e with $I(e_j) \cap S_m$;
15   Make a copy of the subgraph rooted at the target node of e_j, and make e points to the root of the copy ;
16   Replace the label of e_j by $I(e_j) - S_m$;
17   APPEND(e's target,$(F_1 \in S_1) \wedge \cdots \wedge (F_d \in S_d) \rightarrow \langle dec \rangle$,m+1,i) ;
18 **end**
19 **return**

Algorithm 10: APPEND

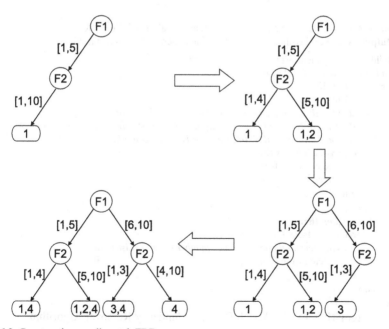

Fig. 6.3 Constructing an all-match FDD

which contains this rule. The all-match list and the residency list for the all-match FDD in Figure 6.2 are in Figure 6.4.

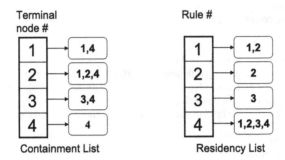

Fig. 6.4 The containment list and the residency List for the all-match FDD in Figure 6.2

The all-match based redundancy removal algorithm works as follows. Given a packet classifier $f : \langle r_1, r_2, \cdots, r_n \rangle$, this algorithm scans f from r_n to r_1, and checks whether each rule is redundant using Theorem 6.1. Whenever a rule is detected as redundant, the rule is removed from f. The pseudocode of the algorithm is shown in Algorithm 11.

Input: A packet classifier $f : \langle r_1, r_2, \cdots, r_n \rangle$, and an all-match FDD t for f.
Output: A packet classifier f' where $f \equiv f'$ and there is no redundant rule in f'.
1 Build the containment list $ConList[1..m]$ from t ;
2 Build the residency list $ResList[1..n]$ from t ;
3 **for** $i := n$ *to* 1 **do**
4   redundant := true ;
5   **for** *each terminal node sequence number tn in ResList[i]* **do**
6   **if** *i is the only value in ConList[tn]) or i is the first value in ConList[tn] and the second value in ConList[tn], say j, satisfies the condition that r_i and r_j have different decisions* **then**
7   redundant := false ;
8   break ;
9   **end**
10   **end**
11   **if** *redundant* **then**
12   remove r_i from f_i ;
13   **for** *each terminal node sequence number tn in ResList[i]* **do**
14   delete i from $ConList[tn]$;
15   **end**
16   **end**
17 **end**

Algorithm 11: All-Match Based Redundancy Removal Algorithm

Consider the packet classifier in Figure 6.1 and its all-match FDD in Figure 6.2. The all-match list and the residency list are in Figure 6.4. We next demonstrate the process of determining whether r_4 is redundant using the all-match based redundancy removal algorithm shown in Figure 11. From the residency list, we know that rule r_4 is contained in terminal nodes $1, 2, 3$ and 4. In terminal node 4, r_4 is the only value, and thus is not redundant. Next, we check whether r_3 is redundant. From the residency list, we know that r_3 is contained in terminal node 3. In terminal node 3, the first value is 3, and is immediately followed by a 4, and r_3 and r_4 have the same decision. According to the Theorem 6.1, r_3 is redundant. Subsequently, we remove r_3 from the packet classifier and delete 3 from the third entry of the all-match list. In a similar fashion, we can further detect that r_2 is redundant and r_1 is not redundant. The resulting packet classifier is shown in Figure 6.5.

$$F_1 \in [1, \ 5] \wedge F_2 \in [1, 10] \ \rightarrow accept$$
$$F_1 \in [1, \ 10] \wedge F_2 \in [1, 10] \rightarrow discard$$

Fig. 6.5 The resulting packet classifier after removing redundant rules from the packet classifier in Figure 6.1

6.2.3 Proof of Complete Redundancy Removal

A packet classifier redundancy removal algorithm is a *complete redundancy removal algorithm* if and only if for any packet classifier the algorithm produces a semantically equivalent packet classifier in which no rule is redundant.

Theorem 6.2. *The All-Match Based Redundancy Removal Algorithm is a complete redundancy removal algorithm.* □

Proof:

Let f be a given packet classifier and let t be an all-match FDD for f. Let f'' be the resulting packet classifier after running the all-match based redundancy removal algorithm. Suppose f'' has a rule r_i that is redundant in f''. Let f' be the resulting packet classifier after the algorithm has examined all the rules from r_{i+1} to r_n and the redundant rules from r_{i+1} to r_n has been removed. Because the algorithm does not remove r_i, r_i is not redundant in f'. According to Theorem 6.1, there is at least a terminal node v that satisfies one of the following conditions:

1. this terminal node only contains i,
2. this terminal node has i as its first value and i is immediately followed by another value j such that r_i and r_j have different decisions.

If v satisfies one of the two conditions, then v still satisfies that condition after the algorithm removes all the redundant rules above r_i, because i will never be deleted from v according to the algorithm. Therefore, r_i is not redundant in f'' according to Theorem 6.1.   □

It is worth noting that the order from n to 1 in detecting redundant rules is critical. If we choose another order, the algorithm may not be able to guarantee complete redundancy removal. Take the order from 1 to n as an example. When we check whether r_i is redundant, suppose r_i is not redundant because there is one and only one terminal node in the all-match FDD that has i as its first value and i is immediately followed by another value j such that r_i and r_j have different decisions. We further suppose j is immediately followed by another value k where r_i and r_k have the same decision. After moving all the redundant rules after r_i, j is possibly removed from the terminal node and consequently r_i and r_k become the first two values in the terminal node and they have the same decision. Thus, r_i becomes redundant.

6.3 Optimization Techniques

In this section, we present two optimization techniques, *decision chaining* and *isomorphic terminal nodes elimination*. These two techniques reduce the amount of memory used in constructing all-match trees and henceforth speed up the construction process.

6.3.1 Decision Chaining

The terminal nodes in an all-match tree may consume a substantial amount of memory. Whenever a subtree in an all-match tree needs to be copied, a substantial amount of time maybe needed to copy terminal nodes, where each terminal node stores an array of rule indexes. Next, we present an optimization technique called *decision chaining*, which stabilized both the amount of memory needed for storing terminal nodes and the time needed for copying subtrees.

Decision chaining is based upon the observation that rules are always added to the end of each terminal node's array. With this observation, we store the rules for each decision in a singly linked list. Furthermore, we order our lists in reverse order so that we can insert rules at the head of the list in constant time. Storing decisions in a singly linked list allows each terminal node to be copied in constant time since we only have to copy the head node of the list. With this optimization, we can reduce that total amount of memory used to store rules since rules can be shared among terminal nodes.

The pseudocode for the modified append subroutine, which is needed to preform this optimization, is shown in Algorithm 12. Figure 6.6 shows the process of constructing an all-match tree with the optimization of decision chaining. Note that r_1

gets reused in the second insertion, and this reuse is maintained for the life of the tree.

Input: A vertex v, a rule $(F_1 \in S_1) \wedge \cdots \wedge (F_d \in S_d) \rightarrow \langle dec \rangle$, a depth m, and a rule number i.
Output: v includes the rule $(F_1 \in S_1) \wedge \cdots \wedge (F_d \in S_d) \rightarrow \langle dec \rangle$ in its all-match structure.

1   Let $F(v) = F_m$ and $E(v) = \{e_1, \cdots, e_k\}$;
2   **if** $m = d + 1$ **then**
3   Add i to the head of v's label list ;
4   **return**
5   **end**
6   **else if** $(S_m - (I(e_1) \cup \cdots \cup I(e_k))) \neq \emptyset$ **then**
7   Add an outgoing edge e_{k+1} with label $S_m - (I(e_1) \cup \cdots \cup I(e_k))$ to v ;
8   Build a decision path from $(F_{m+1} \in S_{m+1}) \wedge \cdots \wedge (F_d \in S_d) \rightarrow \langle dec \rangle$, and make e_{k+1} point to the first node in this path ;
9   Add i to the end of the label of the terminal node of this decision path ;
10   **end**
11   **for** $j := 1$ **to** k **do**
12   **if** $I(e_j) \subseteq S_m$ **then**
13   APPEND(e_j's target,$(F_1 \in S_1) \wedge \cdots \wedge (F_d \in S_d) \rightarrow \langle dec \rangle$,m+1,i) ;
14   **end**
15   Add one outgoing edge e to v, and label e with $I(e_j) \cap S_m$;
16   Make a copy of the subgraph rooted at the target node of e_j, and make e points to the root of the copy ;
17   Replace the label of e_j by $I(e_j) - S_m$;
18   APPEND(e's target,$(F_1 \in S_1) \wedge \cdots \wedge (F_d \in S_d) \rightarrow \langle dec \rangle$,m+1,i) ;
19   **return**
20   **end**

Algorithm 12: APPEND with decision chaining

Decision chaining is compatible and complementary to lazy copying [Meiners et al(2007)Meiners, Liu, and Torng], another optimization technique that we use in constructing all-match trees. Lazy copying can reduce the number of subtrees that are copied during construction.

With decision chaining, we construct an all-match tree by inserting the rules in a classifier from the last to the first. This all-match tree constructing method does not require modification to our redundancy removal algorithm. In addition, when compared to inserting rules from the first to the last, this method can significantly improve the efficiency of lazy copying and decision chaining. This advantage is illustrated in Figure 6.7, where Figure 6.7(a) shows the all-match tree constructed by inserting the rules in the classifier in Figure 6.1 from the first to the last and Figure 6.7(b) shows the all-match tree constructed by inserting the rules in the same classifier from the last to the first. Note that in figure Figure 6.7(a) rule r_4 is replicated across all four terminal nodes and in figure Figure 6.7(b) rule r_4 is stored only in one terminal node. The intuition behind this optimization is that the rules that appear later in a classifier tend to cover larger areas, and thus, they typically intersect with more rules than rules that appear towards the front of the classifier. Inserting a rule that intersects with a large number of rules that have been inserted before it will

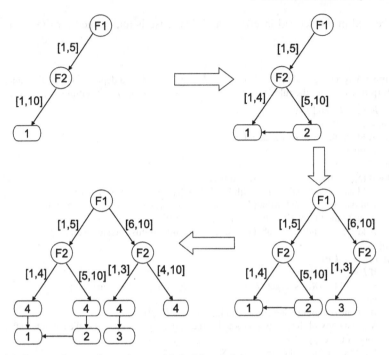

Fig. 6.6 Constructing an all-match tree with decision chaining

result in many subtree being copied. Therefore, inserting rules from back to front results in less subtree copying and more memory sharing.

6.3.2 Isomorphic Terminal Nodes Elimination

In constructing an all-match tree from a packet classifier, the intermediate trees may contain isomorphic nodes. Two nodes v and v' in an all-match tree are *isomorphic* if and only if v and v' satisfy one of the following two conditions: (1) both v and v' are terminal nodes with identical labels; (2) both v and v' are nonterminal nodes and there is a one-to-one correspondence between the outgoing edges of v and the outgoing edges of v' such that every pair of corresponding edges have identical labels and they both point to the same node.

Combining two isomorphic nodes into one node will clearly reduce memory usage. However, the time and space required for detecting all isomorphic node after appending every rule may over weigh its benefit. Therefore, we propose to eliminate only isomorphic terminal nodes, rather than eliminating all isomorphic nodes, for two reasons. First, isomorphic terminal nodes can be detected much more efficiently than isomorphic nonterminal nodes. Second, isomorphic terminal nodes ap-

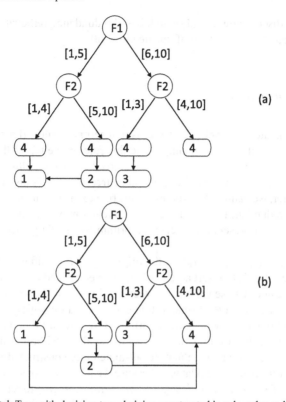

Fig. 6.7 All-Match Tree with decision tree chaining constructed in rule order and reverse order.

pear more often than isomorphic nonterminal nodes, and they consume a significant amount of memory.

We detect isomorphic terminal nodes using collision resistant hash functions. We first iterate over the list of terminal nodes and use the hash function to insert each node into a hash table that store nodes with identical hashes in a hash collision list. During the insertion process, whenever we encounter a hash collision we compare the new node for equality with every node within the hash collision list and add the new node only if it is not equal any node within the list. Once we are finished inserting nodes, we use the hash table entries to construct a list of unique terminal nodes. In our example, the four terminal nodes $\langle 1,4 \rangle$, $\langle 1,2,4 \rangle$, $\langle 3,4 \rangle$, $\langle 4 \rangle$ would all have different hashes so none of the terminal nodes would be eliminated.

6.4 Classifier Incremental Updates

Packet classification rules periodically need to be updated. There are three possible incremental updates on a classifier: inserting a rule, deleting a rule, and modifying a rule. Next, we discuss how our all-match based redundancy removal algorithm can be adapted to handling each type of update efficiently.

6.4.1 Rule Insertion

Typically the update on a packet classifier is to insert new rules. If a rule is inserted on the top of a classifier, we can simply add this rule on the top of the redundancy free classifier produced by our redundancy removal algorithm, although the resulting classifier is not guaranteed to be redundancy free. If a rule is inserted in the middle of a classifier, we cannot directly insert this rule to the redundancy free classifier because some redundant rules that we have removed may become non-redundant. In what follows, we present an efficient algorithm for handing such insertion of a new rule.

Let $f = \langle r_1, \cdots, r_i, r_{i+1}, \cdots, r_n \rangle$ be the original given classifier. Suppose the update on this classifier is to insert a new rule $r_{i'}$ between r_i and r_{i+1}. We let the index i' to be a real number whose value is between i and $i+1$. Next, we present a method for quickly constructing the all-match tree of f', based on which we can use our all-match based redundancy removal algorithm to remove all redundant rules from f'. The basic idea is to insert the new rule $r_{i'}$ to the all-match tree constructed from classifier f. Note that we assume that the all-match tree constructed from classifier f has been kept for handling updates; otherwise we have to reconstruct an all-match tree from the updated classifier $f' = \langle r_1, \cdots, r_i, r_{i'}, r_{i+1}, \cdots, r_n \rangle$ to remove all redundant rules from f'. Since we construct an all-match tree by running updates in sorted order, every terminal node will have the rule numbers in sorted order. We can modify append to perform an insertion by changing how it modifies the terminal node labels. Instead of adding the rule number to the end of the label, the insertion modifies terminal labels by inserting the rule number according to the sorted order. The pseudocode for the insertion function is shown in Algorithm 13. After the all-match tree for $f' = \langle r_1, \cdots, r_i, r_{i'}, r_{i+1}, \cdots, r_n \rangle$ is obtained, we rerun our all-match based redundancy removal algorithm to remove all redundant rules from f'.

If a large number of insertions need to be performed at once, the normal update function can be used; however, before the all-match redundancy removal is rerun, the list of numbers within each terminal node will need to be sorted.

6.4.2 Rule Deletion

For rule deletion, suppose that we delete rule r_i from classifier $f = \langle r_1, \cdots, r_i, r_{i+1}, \cdots, r_n \rangle$, we delete rule r_i from all terminal nodes of the all-match tree constructed from f. The resulting tree is the all-match tree for the updated classifier $f' = \langle r_1, \cdots, r_{i-1}, r_{i+1}, \cdots, r_n \rangle$. Then, we rerun our all-match based redundancy removal algorithm to remove all redundant rules from f'.

Input: vertex v, a rule $(F_1 \in S_1) \wedge \cdots \wedge (F_d \in S_d) \to \langle dec \rangle$, a depth m, and a rule number i.
Output: v includes the rule $(F_1 \in S_1) \wedge \cdots \wedge (F_d \in S_d) \to \langle dec \rangle$ in its all-match structure.

```
/* F(v) = Fm and E(v) = {e1,··· ,ek}                                              */
 1 if m = d + 1 then
 2      Add i to the end of v's label ;
 3      return
 4 end
 5 else if (Sm − (I(e1) ∪ ··· ∪ I(ek))) ≠ ∅ then
 6      Add an outgoing edge ek+1 with label Sm − (I(e1) ∪ ··· ∪ I(ek)) to v ;
 7      Construct a decision path from (Fm+1 ∈ Sm+1) ∧ ··· ∧ (Fd ∈ Sd) → ⟨dec⟩, and make ek+1
        point to the first node in this path ;
 8      Insert i according the sorted order of the label of the terminal node of this decision path ;
 9 end
10 for j := 1 to k do
11      if I(ej) ⊆ Sm then
12          INSERT(ej's target,(F1 ∈ S1) ∧ ··· ∧ (Fd ∈ Sd) → ⟨dec⟩,m+1,i) ;
13      end
14      Add one outgoing edge e to v, and label e with I(ej) ∩ Sm ;
15      Make a copy of the subgraph rooted at the target node of ej, and make e points to the
        root of the copy ;
16      Replace the label of ej by I(ej) − Sm ;
17      INSERT(e's target,(F1 ∈ S1) ∧ ··· ∧ (Fd ∈ Sd) → ⟨dec⟩,m+1,i) ;
18      return
19 end
```

Algorithm 13: INSERT

6.4.3 Rule Modification

For rule modification, we model it as a rule deletion plus a rule insertion. Suppose that we modify rule r_i to be $r_{i'}$ in classifier $f = \langle r_1, \cdots, r_i, r_{i+1}, \cdots, r_n \rangle$. We first get the all-match tree for the classifier $f' = \langle r_1, \cdots, r_{i-1}, r_{i+1}, \cdots, r_n \rangle$ in the way that we handle rule deletions. Second, we obtain the all-match tree for the classifier $f'' = \langle r_1, \cdots, r_{i-1}, r_{i'}, r_{i+1}, \cdots, r_n \rangle$ in the way that we handle rule insertion. Finally, we rerun our all-match based redundancy removal algorithm to remove all redundant rules from f''.

6.5 Redundancy Analysis

In this section, we first introduce the concept of redundancy analysis, the goal of which is to calculate the set of rules that causes a particular rule to be redundant. Then, we present an algorithm for redundancy analysis.

Redundancy analysis is very useful for system administrator to investigate the reason of a rule being redundant. Updates to rule lists often unintentional cause older rules to become redundant. Redundant rules also can indicate misconfiguration within rule lists. For example, in the rule list in Figure 6.1, we have two redundant rules, r_2 and r_3. For each of these rules, we can find rules that make them redundant: r_2 is redundant because r_1 precedes it, and r_3 is redundant because r_4 makes it ineffective.

Formally we define a *redundancy cause set of* r_i to be a set of *causal sets* where a causal set S is a set of rules such that removing all the rules in S will make r_i non-redundant. In our example, the redundancy cause set of r_2 is $\{\{r_1\}\}$ where $\{r_1\}$ is a causal set, and the redundancy cause set of r_3 is $\{\{r_4\}\}$ where $\{r_4\}$ is a causal set.

We observe that the containment and residency lists in Figure 6.4 provide complete information for calculating the causal sets for any rule in a rule list. Recall from Theorem 6.1 that a r_i is redundant if none of its occurrences in the containment list's terminal nodes appear at the front of a list or all of its occurrences at the front of a terminal node list are followed by a rule with the same decision. Therefore, each terminal node that contains r_i provides a cause for r_i's redundancy and forms a causal set.

The redundancy cause sets provide a succinct list of causes for each rule's redundancy, and we compute the redundancy cause set for r_i in the following manner: For each entry j in the residency list, the algorithm enumerates out a single causal set for r_i at the terminal node j in the containment list. According to Theorem 6.1, removing any one set of rules will prevent r_i from becoming redundant. Algorithm 14 contains the psuedocode for computing the redundancy cause set of r_i given a containment and residency list.

Input: A rule number r_i, a containment list $C[1 \ldots m]$, and a residency list $R[1 \ldots n]$.
Output: S is the redundancy cause set of r_i.

1 $S \leftarrow \emptyset$;
2 **for** *each* $j \in R[i]$ **do**
3   *ForeCause* $\leftarrow$ the set of all rules preceding r_i in $C[j]$;
4   *AftCause* $\leftarrow$ the set of all rules between r_i and the first rule that follows it with a
        different decision in $C[j]$;
5   **if** $(ForeCause \cup AftCause) \neq \emptyset$ **then**
6   Add $(ForeCause \cup AftCause)$ to S ;
7   **end**
8 **end**
9 **return** S

Algorithm 14: Find redundancy cause set of r_i

Each causal set can be reviewed by a system administrator to determine if the cause is legitimate or not. Additional techniques can be used to further prune this list of causal set to assist the administrator. For example, we can eliminate causal sets that contain only rules that are older than r_i; this filtering allows the administrator to review the causal sets which have been changed or created during a recent rule list update. Conversely, by eliminating causal sets that contain only rules that are newer than r_i, the administrator can review the original causal sets that r_i had with the classifier when it was originally added. By examining both types of filtered causal sets, the administrator can determine if the intent of r_i is still being enforced.

Part II
New Architectural Approaches

New architectural approaches seek to modify how the TCAM based packet classifiers operate in order to improve efficiency. We propose two approaches: sequential decomposition and topological transformation. Sequential decomposition decomposes a single d-field packet classification TCAM lookup into a sequence of d 1-field TCAM lookups. Topological transformations provide methods to translate the domain of each packet field into a more efficient representation. Both techniques allow for the efficient utilization of TCAM space. These techniques mitigate the effects of range; however, they have the unique advantage that they find optimizations beyond range expansion. This advantage allows for sublinear compression.

Chapter 7
Sequential Decomposition

The problem that sequential decomposition tries to solve in this chapter can be stated intuitively as follows: *how can we squeeze the most information possible into TCAM chips with little or no performance loss?* Solving this problem helps to address almost all limitations of TCAMs. Given a packet classifier, if we can represent it using fewer bits, we can use a smaller TCAM chip, which will result in lower power consumption, less heat generation, less board space, and lower hardware cost. Furthermore, reducing the number of TCAM rules results in less power consumption and heat generation because the energy consumed by a TCAM grows linearly with the number of ternary rules it stores [Yu et al(2005)Yu, Lakshman, Motoyama, and Katz].

Sequential decomposition is composed of four algorithmic approaches to rethinking and redesigning TCAM-based packet classification systems: multi-lookup, pipelined-lookup, packing, and table consolidation. These approaches move beyond the traditional paradigm that performs a single lookup on a single TCAM for each search key. The following two observations of *information redundancy* and a *ternary search key*, which have mostly been ignored in prior work, form the theoretical basis for our new approaches.

Information Redundancy

Information stored in TCAMs tends to have high redundancy from an information theory perspective. Specifically, we observe that the same ternary string for a specific field may be repetitively stored in multiple TCAM entries. For example, in the simple two-dimensional packet classifier in Figure 7.1(a), the strings 001, 010, and 100 from the first field are each stored three times in the TCAM, and the strings 001, 010, and *** from the second field are each stored three times in the TCAM as well. Such information redundancy is primarily due to the multi-dimensional nature of packet classification rules. One source of information redundancy is range expansion in two or more fields.

C. R. Meiners et al., *Hardware Based Packet Classification for High Speed Internet Routers,*   75
DOI 10.1007/978-1-4419-6700-8_7, © Springer Science+Business Media, LLC 2010

Single-lookup		Multi-lookup	
		t_1	
001,001	accept	001	t_2
001,010	accept w. log	010	t_2
001,***	discard	100	t_2
010,001	accept	***	t_3
010,010	accept w. log		
010,***	discard	t_2	
100,001	accept	001	accept
100,010	accept w. log	010	accept w. log
100,***	discard	***	discard
,	discard w. log		
		t_3	
		***	discard w. log

(a)	(b)

Fig. 7.1 Reducing information redundancy

Ternary Search Key

A TCAM chip typically has a built-in Global Mask Register (GMR) that supports
ternary search keys. The GMR of a TCAM chip contains a bit mask that specifies
which bit columns in the chip participate in a search. For example, in a TCAM chip
that contains two entries 1010 and 0100, if the search key is 0101 and the GMR
specifies that only the first two columns participate in this search, the TCAM chip
will return that the lookup key matches the second entry 0100. In essence, the GMR
allows the user to specify the search key in ternary format. In the above example, the
GMR transforms the search key 0101 into 01**. The GMR opens new opportunities
for further improving TCAM space efficiency. Intuitively, the GMR allows multiple
lookup tables to be packed into one TCAM chip where the GMR can be used at run
time to dynamically select the right table to search.

The multi-lookup approach is based on three key observations. First, breaking a
multi-dimensional packet classifier into multiple one-dimensional classifiers greatly
reduces information redundancy in TCAMs. Second, multiple lookup tables can co-
reside in TCAM as long as extra bits are set to distinguish them. Third, a search
key can be segmented into multiple search keys where each is searched in a one-
dimensional classifier. Breaking the two-dimensional classifier in Figure 7.1(a) results in the three one-dimensional classifiers in Figure 7.1(b). Although in this particular example, the number of entries in the original single-lookup table is only two
more than the total number of entries in the multi-lookup tables, the savings will in-
crese significantly as the repetition in each field increases. Furthermore, note that
the width of the multi-lookup tables is much smaller than that of the single lookup
table.

The space efficiency achieved by the multi-lookup approach comes with the price
of more clock cycles to perform each search. Our pipelined-lookup approach speeds
up the multi-lookup approach by pipelining the multiple lookups using multiple
TCAM chips. Interestingly, the pipelined-lookup approach achieves even higher

packet classification throughput than the traditional single-lookup approach because the narrower TCAM entries now fit on the data bus.

The packing approach is based on the following three observations. First, TCAM chips have limited configurability on their width. This prevents us from configuring the TCAM width to exactly the table width, which could cause a significant number of bits in each TCAM entry to be unused. Second, the multi-lookup and pipelined-lookup approaches produce "thin" tables of varying width. Third, search keys for TCAM chips can be ternary, which allows TCAM columns to be dynamically selected for each lookup. The basic idea of the packing approach is that multiple tables can be placed within the same TCAM entries. These tables will be distinguished by the GMR.

In addition to the packing approach, *table consolidation* allows one TCAM table to store multiple classifiers efficiently at the expense of extra SRAM. Table consolidation is based on the two observations. First, TCAM is far more expensive and consumes much more power than SRAM; it makes sense to use a large SRAM with a small TCAM rather than a small SRAM with a large TCAM. Second, semantically different TCAM tables may share common entries, and transferring this redundancy to SRAM removes the information redundancy from the more expensive TCAM.

7.1 Multi-Lookup Approach

Prior work and current practice have assumed the use of a single-lookup for TCAM based systems. We observe that relaxing this assumption could yield unexpected savings on TCAM space with minor throughput degradation. In this section, we propose a multi-lookup approach to redesigning TCAM based systems. We present two algorithms to support this approach: an algorithm for constructing a multi-lookup table and an algorithm for processing packets.

7.1.1 Constructing Multi-lookup Table

The algorithm for constructing a multi-lookup TCAM table from a given packet classifier consists of the following four steps, which are illustrated in Figure 7.2 and Figure 7.3: (1) *FDD Construction*: Constructing a tree-like representation, called a *Firewall Decision Diagram* (FDD), of the packet classifier. (2) *FDD Reduction*: Reducing the size of the FDD. (3) *Table Generation*: Generating a TCAM table from each nonterminal node in the reduced FDD. (4) *Table Mergence*: Merging the generated TCAM tables into a single multi-lookup TCAM table.

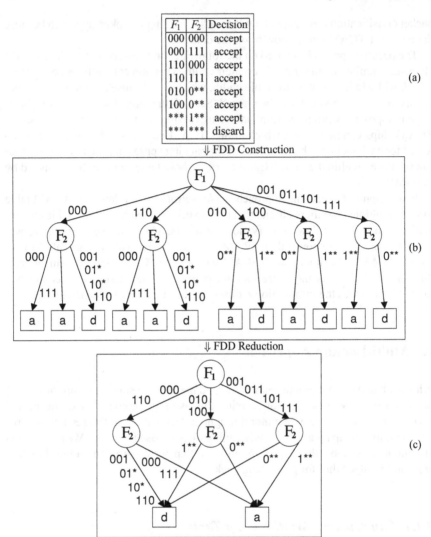

Fig. 7.2 FDD generation and reduction

7.1.1.1 FDD Construction

To generate a multi-lookup TCAM table, we first convert a given packet classifier
to an equivalent firewall decision diagram.Figure 7.2(b) shows the FDD constructed
from the packet classifier in Figure 7.2(a).

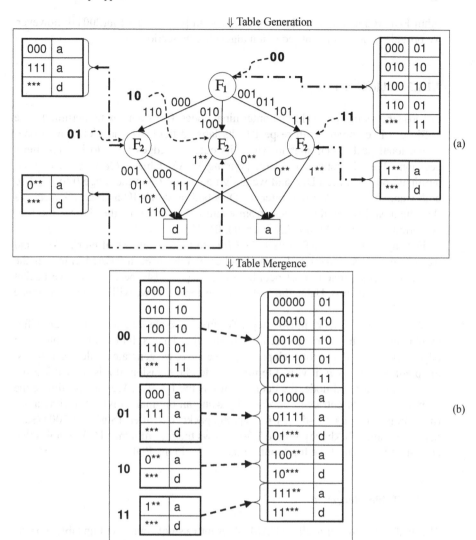

Fig. 7.3 The Multi-lookup scheme

7.1.1.2 FDD Reduction

Reduction is an important step in reducing the total number of TCAM entries in the final multi-lookup table because the reduction step reduces the number of nonterminal nodes, which consequently reduces the number of TCAM entries generated. Figure 7.2(c) shows the resultant FDD after FDD reduction. A brute force algorithm FDD reduction algorithm can be found in [Gouda and Liu(2004)]; however, we provide a more efficient reduction algorithm in Section 7.6.

7.1.1.3 Table Generation

Suppose the reduced FDD has n nonterminal nodes. Consider any nonterminal node v. Since v is complete with respect to its labeled field, we can view v as a one-dimensional packet classifier in which its outgoing edges point to its classifier's decisions. We will construct a corresponding TCAM table $Table(v)$ for each nonterminal node v in the FDD, and we assign a unique ID in the range 0 to $n-1$ to both v and $Table(v)$. We will refer to ID as both node v's ID and table $Table(v)$'s ID. The meaning should be clear from context. For example, the IDs for the four nonterminal nodes in Figure 7.3(a) are 00, 01, 10, and 11.

For any table t, we define its *height* $h(t)$ to be the number of entries in t, and we define its *width* $w(t)$ to be the number of bits in each TCAM entry. In the single lookup approach, most people assume $w(t) = 144$ because the five packet fields require 104 bits. Using the multi-lookup approach, we will show we can make $w(t) = 72$.

We generate $Table(v)$ in two steps. We first generate a correct packet classifier by making one entry for each prefix on each edge. That is, for each of v's outgoing edges e from v to v' and for each prefix p on edge e, we generate a rule r as follows: the predicate of r is p; if v' is a terminal node, the decision of r is the label of v'; if v' is a nonterminal node, the decision of r is the ID of v'. We then minimize the number of TCAM entries in $Table(v)$ by using an optimal, polynomial-time algorithm for minimizing one-dimensional prefix packet classifiers [Suri et al(2003)Suri, Sandholm, and Warkhede]. Figure 7.3(a) shows the four minimal TCAM tables that correspond to each of the four nonterminal nodes in the FDD.

7.1.1.4 Table Mergence

The final step is to merge all n TCAM tables into a single multi-lookup table. For every nonterminal node v, we prepend v's ID to the predicate of each rule in $Table(v)$. Since each table ID provides a unique signature that distinguishes that table's entries from all other table entries, all n tables can be merged into a single multi-lookup table. Figure 7.3(b) shows the resultant multi-lookup table from merging the four TCAM tables in Figure 7.3(a).

7.1.2 Packet Processing

After the multi-lookup TCAM table is built for a d-dimensional packet classifier, the decision for a d-dimensional packet $(p_1, \ldots, p_d)$ can be found by d searches on the

TCAM. The first search key k_1 is formed by concatenating the root node's ID and p_1. Let $f(k_1)$ denote the search result of k_1. The second search key, k_2 is formed by concatenating $f(k_1)$ and p_2. This process continues until we compute $f(k_d)$, which is the decision for the packet. For example, given the two dimensional multi-lookup table in Figure 7.2(e) and a packet $(010, 001)$, the first search key is 00010, which returns 10. The second search key is 10010, which returns *accept* as the decision for the packet.

7.1.3 Analysis

We analyze the impact of the multi-lookup approach on TCAM space and packet classification throughput.

7.1.3.1 Space

We define the space used by a packet classifier in a TCAM chip as the number of classifier entries or rules multiplied by the width of the TCAM chip in bits:

$$space = \# \ of \ entries \times TCAM \ width$$

Given a packet classifier f, let $Single(f)$ denote the resulting single lookup TCAM table and let $Multi(f)$ denote the resulting multi-lookup TCAM table. It follows that width $w(Single(f)) = 144$ because the table must accommodate the 104 bits in the five packet fields, and the number of entries is $h(Single(f))$. Thus, the number of bits required by the single lookup approach is $h(Single(f)) \times 144$. On the other hand, we can safely set width $w(Multi(f)) = 72$. This follows as the maximum width of the five packet fields is 32, which leaves 40 bits for storing a table ID and optionally, the decision. This is more than sufficient for any realistic TCAM for the forseeable future. Thus, the number of bits required by the multi-lookup approach is $h(Multi(f)) \times 72$. The multi-lookup table starts with a 50% reduction in width.

7.1.3.2 Throughput

Based on the above analysis that there are at least 40 bits to store table IDs plus the decision, there is sufficient space to store the decision for each rule in the TCAM entry and still have each TCAM entry fit within the 72 bits of the typical TCAM bus width. Thus, it will require two bus cycles to process each packet field: one cycle to send the search key and one cycle to perform the search and return the result. Given there are five packet fields that need to be processed, the total packet processing time will require ten TCAM bus cycles. The single lookup approach requires either four TCAM bus cycles or five TCAM bus cycles to find the decision for a packet:

four bus cycles if the decision is stored in TCAM, five bus cycles if the decision is stored in SRAM. Note that the overall packet processing throughput for the multi-lookup approach may actually be closer to the packet processing throughput of the single lookup approach because TCAM lookup is normally not the bottleneck of such systems; instead other operations such as moving a packet in and out of queues are the real bottlenecks, so taking a few more bus cycles to process a packet may not have a significant impact on throughput.

7.2 Pipelined-lookup Approach

The multi-lookup approach is an effective method for reducing TCAM space needed for packet classifiers. However, this reduction in space reduces packet classification throughput by requiring multiple lookups on a single TCAM chip. In this section, we present our pipelined-lookup approach, which improves packet throughput by using one TCAM chip for each field. That is, we will use five TCAM chips where, for $1 \leq i \leq 5$, chip i stores table t_i which is the merger of all tables of F_i nodes. Having one merged table per field in a separate TCAM chip enables us to pipeline the multiple lookups needed for processing each packet. Surprisingly, the pipelined-lookup approach can be four to five times faster than the traditional single-lookup approach. Furthermore, separating the tables from different fields yields new opportunities to save bits. The result is that while more TCAM chips are needed, the pipelined-lookup approach can be even more space efficient than the multi-lookup approach. Next, we present the technical details of the pipelined-lookup approach to redesigning TCAM based systems. In particular, we present two algorithms to support this approach: an algorithm for constructing a sequence of d TCAM tables and an algorithm for processing packets.

7.2.1 Pipelined-Table Construction

Our algorithm for constructing a sequence of d tables t_1 through t_5 consists of four steps. The first two steps are FDD construction and FDD reduction, which are similar to the first two steps in the multi-lookup approach. The last two steps, *table generation* and *table mergence*, require some modifications as described below.

7.2.1.1 Table Generation

This step differs from the table generation step in the multi-lookup approach in assigning node IDs in the constructed FDD. Here, we assign each node an ID that can uniquely discriminate that node from all other nodes of the same field. Let m_i be the number of nodes with label F_i in the constructed FDD. The ID assigned to each F_i node consists of $\lceil \log m_i \rceil$ bits. For example, the IDs of the three F_2 nodes in Figure 7.4(a) are 00, 01, and 10. In contrast, in the table generation step of the multi-lookup approach, the ID assigned to each node with label F_i consists of $\lceil \log (\sum_{i=1}^{d} m_i) \rceil$ bits. Note that each F_i node has a unique ID in the context of F_i. We also observe that for field F_1, there will always be a single table. Therefore, we do not need an ID to distinguish this table from tables of other fields. In the remainder of this section, we assume no ID is needed for the F_1 table.

7.2.1.2 Table Mergence

Similar to the table mergence step in the multi-lookup approach, for every nontermi-nal node v, we first prepend v's ID to each rule in $Table(v)$. Second, for every field F_i, we combine all tables of F_i nodes into one table t_i. For example, Figure 7.4(b) shows the two pipelined TCAM tables generated from the FDD in Figure 7.4(a).

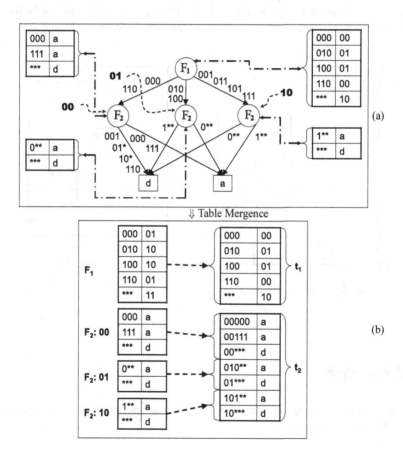

Fig. 7.4 Table generation and table mergence in the pipelined-lookup approach

7.2.2 Packet Processing

Similar to the multi-lookup approach, in the pipelined-lookup approach, a d-dimensional packet search is separated into d searches; however, with the pipelined-lookup approach, these searches can be pipelined. That is, the d TCAM chips are chained together into a pipeline such that the search result of the i-th chip is part of the search key for the $(i+1)$-th chip, and the result of the last chip is the decision for the packet. With such a chain, d packets can be processed in parallel in the pipeline.

Figure 7.5 illustrates the packet processing algorithm for the two tables t_1 and t_2 in Figure 7.4(b). Suppose two packets $(010,001)$ and $(111,010)$ arrive one after the other. When $(010,001)$ arrives, the first search key, 010, is formed and sent to t_1 while the rest of the the packet (001) is forwarded to t_2. When the next packet $(111,010)$ arrives, table t_1 has sent the search result 01 to table t_2. When the first search key for the second packet 111 is formed, the second search key for the first packet 01001 is formed in parallel, and both are sent to tables t_1 and t_2, respectively. This cycle will yield a result of accept for the first packet, and a result of 10 for the second packet. The above process continues for every received packet.

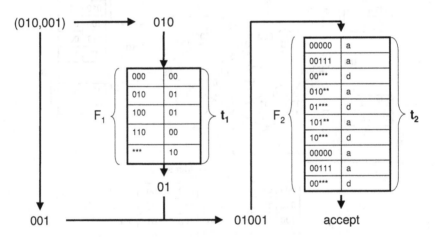

Fig. 7.5 Example of a pipelined-lookup

7.2.3 Analysis

We next analyze the impact of the pipelined-lookup approach on TCAM space and classification throughput.

7.2.3.1 Space

The pipelined-lookup approach is at least as space efficient as the multi-lookup approach, and there are many cases where it uses even fewer TCAM bits. We first observe that the pipelined-lookup approach generates the same number of TCAM entries as the multi-lookup approach. That is, if t is the table formed by the multi-lookup approach and t_1 through t_5 are the tables formed by the pipelined-lookup approach, we have that $h(t) = \sum_{i=1}^{5} h(t_i)$. The space savings of the pipelined-lookup approach results from requiring fewer bits per entry. The first opportunity for saving space comes from the fact that the number of bits needed to encode a table ID in the pipelined-lookup approach is less than or equal to that for the multi-lookup approach as we only need to distinguish a table from other tables with the same field label. The second opportunity for savings comes from the three fields with width 16 or 8 bits. For these fields, we may use width $w(t_i) = 36$ whereas $w(t) = 72$. Specifically, for the source port and destination port fields, we have $36 - 16 = 20$ bits to represent a table ID and optionally the decision of each rule. For the protocol field, we have $36 - 8 = 28$ bits for this purpose. For most classifiers, these bits should suffice. In summary, the pipelined-lookup approach is at least as space efficient as its multi-lookup counterpart; furthermore, there are cases where it is more space efficient.

7.2.3.2 Throughput

The pipelined-lookup approach clearly leads to higher throughput than the multi-lookup approach. More so, the pipelined-lookup approach actually achieves four or five times higher throughput than the traditional single-lookup approach. In the pipelined-lookup approach, because a search key can always be transmitted over the bus in one cycle, the packet classification throughput is one packet per cycle. In contrast, the packet classification throughput of the traditional single-lookup approach is one packet per four or five cycles.

7.3 Packing Approach

In this section, we present the TCAM packing approach. This approach reduces space consumption by allowing multiple rules from different TCAM tables to co-reside in the same TCAM entry. The TCAM packing approach is orthogonal to the multi-lookup and pipelined-lookup approaches in that it can be combined with the two approaches to further improve TCAM space efficiency. The TCAM packing idea is based on the following three key observations:

First, the reconfigurability of TCAM widths is limited. TCAM chips typically only allow entry widths of 36, 72, 144, or 288 bits. This leads to wasted space as typical TCAM tables rarely can be configured to exactly one of these widths. In the standard single lookup approach, up to 40 bits might be unused given the predicate will be 104 bits.

Second, the multi-lookup and pipelined-lookup approaches produce "thin" tables of varying widths. We say the tables are thin because each table focuses on a single packet field. Thus, the table widths are much smaller than 104 bits. The widths vary because the packet fields have different lengths: 8, 16, and 32, and these predicate bits form a significant fraction of each table entry. Having multiple fields of varying widths provides opportunities to better approach the standard TCAM widths.

Third, the search key for TCAM chips can be ternary. In other words, TCAM columns are dynamically selectable for each lookup. A typical TCAM chip has a *global mask register*, which dynamically selects the columns that participate in a lookup. The global mask register allows multiple entries from different lookup tables to co-reside in the same TCAM entry without conflicting with each other.

We developed two TCAM packing schemes, which we call *strict partitioning* and *shadow packing*.

7.3.1 Strict Partitioning

7.3.1.1 Basic Strict Partitioning

The basic idea of strict partitioning is to divide a TCAM chip into multiple columns and distribute multi-lookup or pipelined-lookup tables among these columns. The distribution needs to satisfy the following two conditions: (1) all tables in the same column must have different node IDs, (2) all the rules in a table are stored in only one column. Multiple tables in the same column are discriminated by their table ID. Multiple columns in the same TCAM chip are discriminated by the GMR of the chip. The appropriate GMR can be selected by using a *column ID* which must have enough bits to discriminate all the columns.

We illustrate the strict partitioning scheme using the four multi-lookup tables in Figure 7.2(d) in a TCAM chip with entry width 21 bits. Figure 7.6 shows a possible arrangement of the four tables: table 00 in column 1, table 01 in column 2, and tables 10 and 11 in column 3. We use column IDs 00, 01, and 10 for columns 1, 2, and 3, respectively. We decode the first entry 00 000 01@01 in column 1 as follows. The first two bits 00 encode the table ID, the next three bits 000 are the rule predicate, and the last four bits 01@01 encode the rule decision where the first two bits encode the next table ID and the last two bits encode the column ID of the next table. By partitioning the TCAM chip into three separate columns, the TCAM chip is essentially divided into three TCAM chips. Lookups in the TCAM chip are performed by padding the search key with the appropriate ternary bits via the GMR. For example, to lookup 111 in table 10 of column 10, the lookup key is ***************10111*.

Storing Decisions: Because packing schemes can use previously unused bits in the multi-lookup and pipelined-lookup approaches to store rules, storing the deci-

TCAM chip		
Col 00	Col 01	Col 10
00 000 01@01	01 000 a	10 0** a
00 010 10@10	01 111 a	10 *** d
00 100 10@10	01 *** d	11 1** a
00 110 01@01		11 *** d
00 *** 11@10		

Fig. 7.6 Strict partitioning of a TCAM chip

sion of each rule in the TCAM entry is not space or cost effective. Therefore, for all our packing schemes, we assume that the rule decisions are stored in SRAM. Figure 7.7 shows a version of the strict partitioning in Figure 7.6 with decisions stored in SRAM.

TCAM chip				SRAM Table		
Col 00	Col 01	Col 10		Col 1	Col 2	Col 3
00 000	01 000	10 0**		01@01	a	a
00 010	01 111	10 ***		10@10	a	d
00 100	01 ***	11 1**		10@10	d	a
00 110		11 ***		01@01		d
00 ***				11@10		

Fig. 7.7 Decisions in SRAM

Packing multi-lookup tables vs. packing pipelined-lookup tables: Recall that all our packing schemes can be applied to both the multi-lookup and pipelined-lookup approaches. The only difference is that when we pack multi-lookup tables into one TCAM chip, we need to deal with tables of variable width. In contrast, when we pack pipelined-lookup tables into d TCAM chips, we only deal with tables of the same width for each chip, ignoring minor differences induced by table IDs.

Reassigning Table IDs With Fewer Bits: The original table IDs were used to distinguish a table from either all other tables (in the multi-lookup approach) or all other tables of the same field (in the pipelined-lookup approach). However, in a packing scheme, we only need to distinguish a table from the other tables in the same column. Therefore, we can often use fewer bits for tables IDs. In our packing schemes, after tables are allocated to columns, we reassign table IDs using the least number of needed bits, and the decisions for the rules have to be updated to reflect the new table IDs. Note that different columns may have different table ID widths, and rule decisions may have different lengths. In the strict partitioning scheme, for a column with n tables, the number of bits in the reassigned ID of each table is $\lceil \log n \rceil$. Figure 7.8 shows a version of strict partitioning in Figure 7.7 with table IDs reassigned with fewer bits.

Processing Packets: We describe the algorithm for processing packets under the strict partitioning approach using examples. Suppose the given packet is $(000, 110)$, the first TCAM lookup is 000******* and the lookup result is the index value of

TCAM chip				Decision Table		
Col 00	Col 01	Col 10		Col 00	Col 01	Col 10
000	000	0 0**		@01	a	a
010	111	0 ***		0@10	a	d
100	***	1 1**		0@10	d	a
110		1 ***		@01		d
***				1@10		

Fig. 7.8 Reassigning the table IDs

0. This index value is used to find entry 0 in the column 00 in the SRAM to find the decision of @01, which means that the second lookup should be performed on column 01. To further perform the second lookup, the GMR is modifed to make the second lookup key ***110****. The result of the second lookup is the index value of 1, which means the decision is stored in the second entry of column 01 in SRAM. The second entry of column 01 in SRAM is "*a*", which means that the final decision for the given packet is *accept*.

7.3.1.2 Optimized Strict Partitioning

Given a set of TCAM tables to be packed in a single TCAM chip, there are many ways to do strict partitioning. First, we can choose the TCAM width to be 36, 72, 144, or 288. Second, for each possible TCAM width, there are many possible ways to divide the TCAM. Third, for each possible division of the TCAM, there are many possible ways to allocate TCAM tables to columns. Among all possible strict partitioning solutions for a TCAM chip, we want to find the solution that uses the least TCAM space under throughput constraints. Note that the classification throughput may decrease as the TCAM width increases. We formally define the *Strict Partitioning Optimization Problem* as follows. We omit throughput constraints in the definition for ease of presentation.

Definition 7.1. Given n TCAM tables $t_1, \ldots, t_n$, find a partition of the tables to m sets $c_1, \ldots, c_m$ that minimizes the objective $TW \times \max_{i=1}^{m} \text{Rules}(c_i)$ such that

- $\bigcup_{i=1}^{m} c_i = \{t_1, \ldots, t_n\}$
- $\sum_{i=1}^{m} (\max_{t_j \in c_i}(\text{w}(t_j)) + \log|c_i|) \leq TW$

where $TW \in \{36, 72, 144, 288\}$, $\text{w}(t_j)$ denotes the width of table t_j, $|t_j|$ denotes the number of rules within t_j, and $\text{Rules}(c_i)$ denotes the total number of table entries in c_i, which is $\sum_{t_j \in c_i} |t_j|$.

The problem of makespan scheduling on multiple identical machines [Garey and Johnson(1978)], which is NP-complete, is a special case of this problem. Thus, the strict partitioning optimization problem is NP-complete. It belongs to NP because the solution can be verified in polynomial time.

7.3.2 Shadow Packing

In strict partitioning, we viewed columns as the primary dimension of TCAM chips and sought to pack tables into fixed width columns. In shadow packing, on the other hand, we view rows as the primary dimension of TCAM chips and seek to pack tables within fixed height rows. We consider shadow packing because of the following two observations.

First, with strict partitioning, when tables of varying width are allocated to the same column, the number of bits assigned to each table t is equal to $h(t) \times w(t')$ where t' is the widest table assigned to that column. This leads to many unused bits if tables of different widths are assigned to the same column. On the other hand, horizontally packed tables can be placed next to each other as keeping the vertical boundaries across multiple tables is unnecessary. Of course, there may be wasted bits if tables of different heights are packed in the same column. We will allow tables to be stacked in the same row if they fit within the row boundaries.

Second, with strict partitioning, the table ID's between tables in different columns cannot be shared. Thus, the number of bits used for table IDs grows essentially linearly with the number of columns. On the other hand, horizontally aligned tables in the same row can potentially share some "row ID" bits in their table IDs; these tables would be distinguished by their horizontal offsets.

Based on the above two observations, we design the shadow packing scheme that achieves more space efficiency by packing tables horizontally and allowing multiple tables to share bits in their IDs. We first define the concept of shadowing for table ID inheritance and the concept of shadow packing trees for representing table ID inheritance. Then, we present the shadow packing algorithm and discuss the procedure for processing packets with shadow packing.

7.3.2.1 Shadowing

In Figure 7.9(a), table t_0 shadows tables t_{00} and t_{01}. We define the concept of shadowing as follows:

Definition 7.2 (Shadowing Relationship). For a table t stored in a TCAM, we use $VBegin(t)$ and $VEnd(t)$ to denote the vertical indexes of the TCAM entries where the table begins and ends respectively, and use $HBegin(t)$ and $HEnd(t)$ to denote horizontal indexes of the TCAM bit columns where the table begins and ends respectively. For any two tables t_1 and t_2 where $[VBegin(t_2), VEnd(t_2)] \subseteq [VBegin(t_1), VEnd(t_1)]$ and $HEnd(t_1) < HBegin(t_2)$, we say t_1 shadows t_2.

When table t_1 shadows t_2, the ID of t_1 can be reused as part of t_2's ID. Suppose table t shadows tables $t_1, \cdots, t_m$, because t's ID defines the vertical TCAM region $[Begin(t), End(t)]$, each t_i ($1 \leq i \leq m$) can use t's ID to distinguish t_i from tables outside $[Begin(t), End(t)]$ vertically, and use $\lceil \log m \rceil$ bits to distinguish t_i from tables inside $[Begin(t), End(t)]$ vertically. Horizontally, table t and each table t_i can be distinguished by properly setting the GMR of the TCAM.

Definition 7.3 (Shadow Packing). Given a region defined vertically by $[v_1, v_2]$ and horizontally by $[h_1, h_2]$, all tables completely contained within this region are *shadow packed* if and only if there exist m $(m \geq 1)$ tables $t_1, \cdots, t_m$ in the region such that the following three conditions hold:

1. $v_1 = VBegin(t_1)$, $VEnd(t_i) + 1 = VBegin(t_{i+1})$ for $1 \leq i \leq m-1$, $VEnd(t_m) \leq v_2$;
2. no tables are allocated to the region defined vertically by $[VEnd(t_m) + 1, v_2]$ and horizontally by $[h_1, h_2]$;
3. for each i $(1 \leq i \leq m)$, the region defined vertically by $[VBegin(t_i), VEnd(t_i)]$ and horizontally by $[HEnd(t_i) + 1, h_2]$ either has no tables or the tables allocated to the region are also shadow packed.

For example, the tables in Figure 7.9(a) are shadow packed. Figure 7.9(b) shows the tree representation of the shadowing relationship among the tables in Figure 7.9(a).

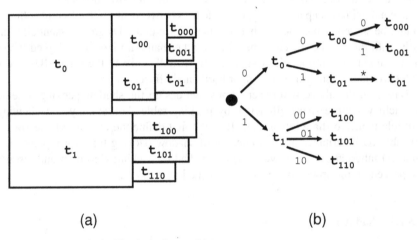

(a)   (b)

Fig. 7.9 Shadow packed tables & shadow packing tree

7.3.2.2 Shadow Packing Algorithm

Given a set of tables and a TCAM region, the shadow packing algorithm allocates the tables into the region. The goal of a shadow packing algorithm is to minimize the number of TCAM entries occupied by the tables, i.e.,, to minimize $VEnd(t_m)$. We call this minimization problem the *Shadow Packing Optimization Problem*. This problem becomes more difficult as we recurse because we must also address which tables should be allocated to which region. Whether this problem can be solved in polynomial time is an open problem.

In this chapter, we present a shadow packing algorithm *SPack*, which has been shown to be effective in our experiments on real-life packet classifiers. The basic idea of SPack is as follows. Given a set of tables S and a TCAM region, SPack first finds the tallest table t that will fit in the region where ties are broken by choosing the fattest table. SPack returns when there are no such tables. Otherwise, SPack places t in the top left corner of the region, and SPack is recursively applied to $S - \{t\}$ in the region to the right of t. After that, let S' be the set of tables in S that have not yet been allocated. SPack is applied to S' in the region below t. Intuitively, SPack greedily packs the tallest (and fattest) possible table horizontally. The pseudocode of SPack is shown in Algorithm 15.

Input: S : a set of tables, and a region $[v_1, v_2], [h_1, h_2]$.
Output: S': the set of tables in S that have not been packed.

1 Find the tallest table $t \in S$ that will fit in $[v_1, v_2], [h_1, h_2]$ such that ties are broken by choosing the fattest table ;
2 **if** *no table is found* **then**
3   **return** S ;
4 **else**
5   Place t in the top left corner of $[v_1, v_2], [h_1, h_2]$;
6      $S'' \leftarrow$ SPack$(S', \text{VBegin}(t), \text{VEnd}(t), \text{HEnd}(t) + 1, h_2)$;
7   **return** SPack$(S'', \text{VEnd}(t) + 1, v_2, h_1, h_2)$;
8 **end**

Algorithm 15: Shadow Packing (SPack)

We, however, must compute the initial SPack region. The height of the initial region is the total number of rules within the set of tables. We do not need to set this value carefully because SPack only moves to another row when all the remaining tables do not fit in any of the current shadows. The width is more complicated and must be computed iteratively. For each valid TCAM width $w \in \{36, 72, 144, 288\}$, we set the initial width to be w and run SPack. Once we have a packing, we determine the number of bits b that are needed for node IDs. If the packing could accommodate these extra b bits, we are done. Otherwise, we choose an aggressive backoff scheme by recursing with a width of $w - b$. It is possible, particularly for $w = 36$, that no solution will be found. To determine which TCAM width we should use, we choose the width $w \in \{36, 72, 144, 288\}$ whose final successful value resulted in the fewest number of entries. Note that there are other possible strategies for determining the width of the SPack regions; for instance, instead of reducing the region width by b, the width could be reduced by 1. Furthermore, to speed up this process, SPack can be modified to abort the packing once it detects that the table packing and IDs can not fit within the region.

Reassignng Table IDs and Rule Decisions: Because shadow packing establishes a hierarchy of table IDs, each table needs a new ID, and all the rule decisions need to be remapped to reflect these new IDs. Each table ID is determined by a tree representation similar to the one found in Figure 7.9(b), which we call a *shadow packing tree*. For each node v in a shadow packing tree, if v has $m > 1$ outgoing edges, each outgoing edge is uniquely labeled using $\lceil \log m \rceil$ bits; if v has only one outgoing edge, that edge is labeled $*$. For each table t, let v be the corresponding node in the shadow packing tree. All the bits along the path from the root to v are all the bits needed to distinguish t from all other tables. Note that the $*$ corresponds to a table where no additional ID bits are needed. In our shadow packing algorithm, we reserve l bit columns in the TCAM where l is the maximum number of bits needed to the distinguish a table. Reserving some bit columns for storing table IDs has the advantage of simplifying the processing of packets since the bit columns containing the table IDs are fixed in the TCAM.

Figure 7.10(a) shows the shadow packing tree for the four tables in Figure 7.2(d) and their reassigned table IDs. Figure 7.10(b) shows the final memory layout in the TCAM chip after shadow packing and the conceptual memory layout of the decision table within SRAM. The one bit ID column in Figure 7.10(b) is needed to distinguish between the tables with original IDs 01 and 11. Note that table 10 shares the table ID 0 with table 01 as it is the only table in table 01's shadow. To make the decision table in Figure 7.10(b) easier to understand, we encode it in a memory inefficient manner using columns.

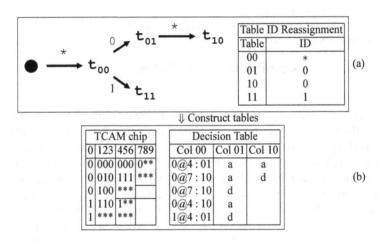

Fig. 7.10 The shadow packing process.

Processing Packets: We describe the algorithm for processing packets under the shadow packing approach using examples. Given a packet $(000, 111)$, the first TCAM lookup is *000******, and the lookup result is the index value of 0. This index value is used to find entry 0 in the column 00 in the SRAM which contains the decision of $0@4 : 01$. The $0@4$ means that the second lookup key should occur in table ID 0 at horizontal offset of 4, and the 01 means that decision of the next search is located in column 01 in SRAM. To perform the second lookup, the GMR is modified to make the second lookup key 0***111***. The result of the second lookup is the index value of 1, and the decision stored in the second entry of column 01 in SRAM is retrieved, which is *accept.*

7.3.3 Strict Partitioning vs. Shadow Packing

We now compare the space efficiency of strict partitioning and shadow packing. The sole advantage of strict partitioning is that it has no horizontal boundaries. On the other hand, shadow packing has two key advantages. It has no vertical boundaries, and tables in the same row can share some table ID bits. Furthermore, we mitigate some of the disadvantage of horizontal boundaries by greedily packing tables in the shadow of other tables.

7.4 Table Consolidation

The basic idea of table consolidation is to use one TCAM table to represent multiple TCAM tables. Table consolidation is motivated by the following two observations. First, two TCAM tables may share common entries, which result in the same information being stored multiple times. Second, existing TCAM-based packet classification systems are based on a "fat" TCAM and "thin" SRAM architecture, which means that the majority of the information (i.e.,, the predicates of rules) representing a packet classifier is stored in TCAMs and little information (i.e.,, the decision of rules) is stored in SRAMs. However, because TCAMs are much more expensive than SRAMs, we ideally would store more information in SRAMs and less information in TCAMs.

We begin with two new concepts: k-decision rule and k-decision classifier. A *k-decision rule* is a classification rule whose decision is an array of k decisions. A *k-decision classifier* is a sequence of k-decision rules following the first-match semantics. We formally define the table consolidation problem as follows:

Definition 7.4 (Table Consolidation Problem). Given k 1-decision classifiers $\mathbb{C}_1, \cdots, \mathbb{C}_k$, find a k-decision classifier $\mathbb{C}$ such that for any i $(1 \leq i \leq k)$, the condition $\mathbb{C}_i \equiv C[i]$ holds.

We emphasize that a k-decision classifier can be viewed as a 1-decision classifier if we view the array of k decisions of each rule as one decision. In general, a k-decision classifier can be viewed as a k'-decision classifier where $k' < k$, if we treat some decisions as one decision.

7.4.1 Table Consolidation Algorithm

We use multi-match FDDs to facilitate table consolidation. A multi-match FDD satisfies all the properties of an all-match FDD except the condition $j \in \{i_1, i_2, \cdots, i_k\}$ in the 5th property.

Our table consolidation algorithm works as follows. First, given a set of k classifiers $\mathbb{C}_1, \cdots, \mathbb{C}_k$, we concatenate the set of classifiers into one classifier $\mathbb{C}_{1-k} =$

$\mathbb{C}_1 | \cdots | \mathbb{C}_k$. Second, we construct a multi-match FDD f from $\mathbb{C}_{1-k}$ such that f satisfies the following additional condition: for any decision path $\mathscr{P}$, the terminal node of $\mathscr{P}$ consists of k numbers $\{m_1, m_2, \cdots, m_k\}$ where, for each i $(1 \leq i \leq k)$, rule r_{m_i} is the first rule in $\mathbb{C}_i$ that contains $\mathscr{P}$. In other words, for any decision path $\mathscr{P}$ in the multi-match FDD f and for any classifier $\mathbb{C}_i$, only the index of the first rule in $\mathbb{C}_i$ that contains $\mathscr{P}$ is included in the label of $\mathscr{P}$'s terminal node. Third, after the multi-match FDD f is constructed, we run the TCAM Razor algorithm presented in [Meiners et al(2007)Meiners, Liu, and Torng] on f and generate the final compact k-decision classifier C. Figure 7.11 shows the process of consolidating two TCAM tables t_1 and t_2. The final 2-decision classifier is TCAM table t_4. The correctness of the table consolidation algorithm is based on Theorem 7.1.

Theorem 7.1. *Given k 1-decision classifiers $\mathbb{C}_1, \cdots, \mathbb{C}_k$, the table consolidation algorithm generates a k-decision classifier $\mathbb{C}$ where for any i $(1 \leq i \leq k)$, the condition $\mathbb{C}_i \equiv C[i]$ holds.*

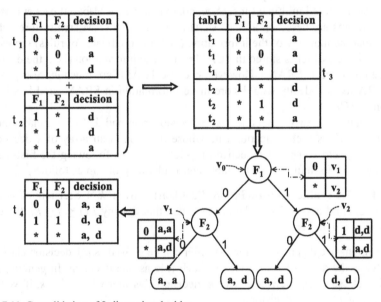

Fig. 7.11 Consolidation of 2-dimensional tables

7.4.2 Hierarchical Table Consolidation

In building a multi-match FDD from a set of classifiers, each classifier influences the shape of the FDD and causes the FDD to grow. To localize the expansion impact of one classifier on others, we propose the following *hierarchical table consolidation* strategy. Given k classifiers, first, we equally divide them into $\lceil k/m \rceil$ buckets where every bucket has m classifiers except for one bucket that may have less than m classifiers. Second, for the classifiers in each bucket, we apply the table consolidation algorithm and get an m-decision classifier. Thus, we get $\lceil k/m \rceil$ classifiers. By treating each m-decision classifier as a 1-decision classifier, we apply the above process again on the $\lceil k/m \rceil$ classifiers. This process repeats until we get the final k-decision classifier. We refer to the strategy of choosing $m = k$ as *flat table consolidation*. Theorem 7.2 establishes the correctness of hierarchical table consolidation.

Theorem 7.2 (Hierarchical Table Consolidation Theorem). *Given the same input of k 1-decision classifiers and the same table consolidation algorithm, the strategies of hierarchical table consolidation and flat table consolidation output the same k-decision classifier.*

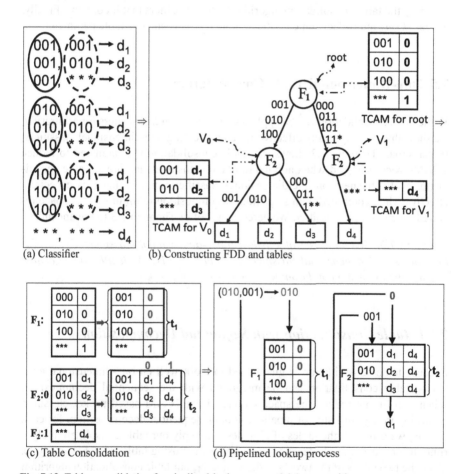

Fig. 7.12 Table consolidation for pipelined-lookup sequential decomposition

7.4.3 TCAM/SRAM Space Tradeoff via Bounded Consolidation

Table consolidation creates a tradeoff between TCAM storage and SRAM storage. On one hand, merging multiple classifiers into one classifier results in less TCAM storage. On the other hand, each entry in the resulting classifier requires more SRAM for storing the decision list. We could simply merge all classifiers into a single classifier; however, this may require more SRAM space than what is available. To address this issue, we propose the following bounded consolidation scheme. The basic idea of bounded consolidation is to limit the number of classifiers that we combine. Given a set of k 1-decision classifiers, we first sort the classifiers in decreasing (or increasing) order according to size (i.e.,, the number of rules in each classifier). Second, we partition the sorted classifiers into $\lceil k/m \rceil$ chunks where the first $\lceil k/m \rceil - 1$ chunks are of uniform size m ($1 \leq m \leq k$). Third, for every chunk, we apply the table consolidation algorithm to the classifiers that it contains. Finally, we get $\lceil k/m \rceil$ multi-decision classifiers. Note that m is an adjustable parameter.

7.5 One-Dimensional Table Consolidation

While table consolidation can be applied to d-dimensional classifiers for arbitrary d, we show that it is especially effective for consolidating one-dimensional classifiers. In particular, Theorem 7.3 shows that table consolidation is guaranteed to reduce TCAM space occupied when applied to one-dimensional classifiers. Table consolidation's effectiveness on one-dimensional classifiers implies that it is especially effective when combined with sequential decomposition to minimize the space required by any single classifier. We defer the proof of Theorem 7.3 to the appendix.

Theorem 7.3. *Given any set of k 1-decision 1-dimensional classifiers $\mathbb{C}_1, \cdots, \mathbb{C}_k$, the k-decision 1-dimensional classifier $\mathbb{C}$ output by the TCAM distillation algorithm satisfies the following condition:* $|C| \leq |C_1| + \cdots + |C_k| - k + 1$.

7.5.1 Table Consolidation with Sequential Decomposition

Because table consolidation is guaranteed to work well on one-dimensional classifiers, table consolidation can be integrated with the sequential decomposition scheme to further reduce the space required by a TCAM-based packet classifier. Basically, in the third step of pipelined-lookup sequential decomposition, for each field F_i, we group all the tables of F_i nodes and apply our table consolidation algorithm to produce a multi-decision table rather than using table ID's. For example, given the two tables of the two F_2 nodes v_0 and v_1, our table consolidation algorithm outputs the 2-decision table t_2 in Figure 7.12(c). In this case, table consolidation reduced the required number of TCAM entries from 4 to 3.

The packet lookup process on the d multi-decision TCAM tables proceeds as follows. The first table has only one column of decisions and each decision is the decision column index of the second table. Similarly, each decision in the i-th ($1 \leq i < d$) table is the decision column index of the next table. We illustrate the packet lookup process using the example in Figure 7.12(d). Given a packet (010,001), we first use 010 to search the table t_1 and get result 0. Second, we use 001 to search table t_2 and get search result (d_1, d_4). The final result is the first element of (d_1, d_4), which is d_1. Note that each TCAM chip has its own SRAM. Therefore, the packet lookup process can be pipelined to achieve the throughput of one packet per cycle.

7.5.2 Coping with More Fields than TCAM chips

If we have fewer TCAM chips than packet fields, we have two choices. One is to strategically combine selected fields into one field until the number of fields is equal to the number of TCAM chips. We can use the FDD structure to facilitate combination of multiple fields. For example, given a d-dimensional classifier over fields $F_1, F_2, \cdots, F_d$ and the corresponding FDD, we can combine the first k ($1 < k < d$) fields into one field by treating the subgraph rooted at each F_{k+1} node as a terminal node and then use our TCAM Razor algorithm presented in [Meiners et al(2007)Meiners, Liu, and Torng] to generate a k-dimensional table for the k-dimensional FDD. Then, for each subgraph rooted at an F_{k+1} node, we can apply the above process recursively.

A second choice is to employ the sequential decomposition multi-lookup approach where we store tables from multiple fields on a single TCAM chip. For example, if we have d dimensions and q chips, we would store tables from either $\lceil d/q \rceil$ or $\lfloor d/q \rfloor$ fields in each chip in the pipeline. Each stage of the pipeline would require either $\lceil d/q \rceil$ or $\lfloor d/q \rfloor$ lookups. This would result in a classification throughput of one packet per $\lceil d/q \rceil$ TCAM bus cycles.

7.6 Implementation Issues

7.6.1 TCAM Update

Packet classification rules periodically need to be updated. The common practice for updating rules is to run two TCAMs in tandem where one TCAM is used while the other is updated [Lekkas(2003)]. All our approaches are compatible with this current practice. Because our algorithms are efficient (running in milliseconds) and the resultant TCAM lookup tables are small, updating TCAM tables can be efficiently performed.

If an application requires very frequent rule update (at a frequency less than a second, for example), we can handle such updates in a batch manner by chaining the TCAM chips in our proposed architecture after a TCAM chip of normal width (144 bits), which we call the "hot" TCAM chip. When a new rule comes, we add the rule to the top of the hot TCAM chip. When a packet comes, we first use the packet as the key to search in the hot chip. If the packet has a match in the hot chip, then the decision of the first matching rule is the decision of the packet. Otherwise, we feed the packet to the TCAM chips in our architecture described as above to find the decision for the packet. Although the lookup on the hot TCAM chip adds a constant delay to per packet latency, the throughput can be much improved by pipelining the hot chip with other TCAM chips. Using batch updating, only when the hot chip is about to fill up, we need to run our topological transformation algorithms to recompute the TCAM lookup tables.

7.6.2 Non-ordered FDDs

Recall that the FDD construction algorithm that we used produces ordered FDDs, that is, in each decision path all fields appear in the same order. However, ordered FDDs may not be the smallest when compared to non-ordered FDDs. The FDD construction algorithm can be easily modified so that different subtrees may use different field ordering. By adding field information to the decisions in each table entry, we can easily accommodate different field orderings. Thus, the packet processing algorithm for both multi-lookup and pipelined-lookup can select the correct field for each lookup. The size advantage of non-ordered FDDs comes at the cost that FDD reduction will not be able to process subtrees that have different field orders. Nevertheless, the use of *non-ordered FDDs* does open new possibilities for further optimizations.

7.6.3 Lookup Short Circuiting

So far, we have assumed the use of *full-length FDDs* where in each decision path all fields appear exactly once. Actually, this constraint can be relaxed so that some paths may omit unnecessary fields when a node in the path contains only one outgoing edge. In this case, the node along with singleton outgoing edge can be pruned. Using FDDs that are not full-length has the advantage of reducing FDD size and consequently reducing the total number of tables. Furthermore, this optimization allows some specific decision paths to be performed with a reduced number of lookups, which will allow for faster packet processing when the tables are processed in a multi-lookup fashion. Therefore, we call this optimization technique *lookup short circuiting*. Similar to the use of non-ordered FDDs, this optimization technique requires storing field information in the decisions.

7.6.4 Best Variable Ordering

In converting a packet classifier to an equivalent FDD, the order of the fields used by decision paths has a significant impact on the size of the resulting FDD. Given that fewer nodes in an FDD normally lead to a smaller multi-lookup or pipelined-lookup table, choosing a good variable order (i.e.,, field order) is important in FDD construction. Given a packet classifier that has five fields, we can easily try all $5! = 120$ permutations to find the best permutation for that particular packet classifier.

Chapter 8
Topological Transformations

One approach for mitigating the effects of range expansion has been to reencode critical ranges. The basic idea is to reencode a given packet and use the reencoded packet as the TCAM search key. For instance, Liu [Liu(2002)], Lunteren and Engbern [van Lunteren and Engbersen(2003)], and Pao et al. [Pao et al(2006)Pao, Li, and Zhou] all proposed methods of representing specific ranges as special bit-strings using extra TCAM bits. Lakshminarayan et al.and Bremler-Barr and Hendler proposed to replace the prefix encoding format with alternative ternary encoding formats, called DIRPE [Lakshminarayanan et al(2005)Lakshminarayanan, Rangarajan, and Venkatachary] and SRGE [Bremler-Barr and Hendler(2007)], respectively.

Previous reencoding schemes suffer from two fundamental limitations. First, they only consider range fields and ignore all other fields; thus, they miss many optimization opportunities that can be applied to prefix fields as well. It was not realized that packet classifiers often have the potential of being minimized in TCAM even when no fields are specified in ranges. Second, they require either computationally or economically expensive reencoding steps that do not easily integrate into existing packet classification systems. As each packet needs to be reencoded before it can be used as a search key, previous range reencoding schemes propose to perform packet reencoding using software, which greatly increases packet processing time, or customized hardware, which is expensive from a design, cost, and implementation perspective.

In this chapter, we take two novel views on range reencoding that are fundamentally different from previous range reencoding schemes. First, we view range reencoding as a topological transformation process from one colored hyperrectangle to another. Whereas previous range reencoding schemes only deal with range fields, we perform reencoding on every packet field. Specifically, we propose two orthogonal, yet composable, reencoding schemes: domain compression and prefix alignment. In domain compression, we transform a given colored hyperrectangle, which represents the semantics of a given classifier, to the smallest possible "equivalent" colored hyperrectangle. In prefix alignment, on the other hand, we strive to transform a colored hyperrectangle to an equivalent "prefix-friendly" colored hyperrectangle where the ranges align well with prefix boundaries, minimizing the costs of range expansion. Second, we view range reencoding as a classification process that can be implemented with small TCAM tables. Thus, while a preprocessing step is still required, it can be easily integrated into existing packet classification systems using the same underlying TCAM technology. Furthermore, implementing our schemes on a pipeline of TCAM chips even increases packet classification throughput because our schemes enable the use of TCAM chips of small width.

Domain Compression: The fundamental observation is that in most packet classifiers, many coordinates (i.e.,, values) within a field domain are equivalent. The idea of domain compression is to reencode the domain so as to eliminate as many redundant coordinates as possible. This type of reduction not only leads to fewer rules, but also narrower rules, which results in smaller TCAM tables. From a geometric perspective, domain compression "squeezes" a colored hyperrectangle as much as possible. For example, consider the colored rectangle in Figure 8.1(A) that represents the classifier in Figure 8.1(H). In field F_1 represented by the X-axis, all values in $[0,7] \cup [66,99]$ are equivalent; that is, for any $y \in F_2$ and any $x_1, x_2 \in [0,7] \cup [66,99]$, packets (x_1, y) and (x_2, y) have the same decision. Therefore, when reencoding F_1, we can map all values in $[0,7] \cup [66,99]$ to a single value, say 0. By identifying such equivalences along all dimensions, the rectangle in Figure 8.1(A) is reencoded to the one in Figure 8.1(D), whose corresponding classifier is shown in Figure 8.1(I). Figures 8.1(B) and (C) show the two transforming tables for F_1 and F_2, respectively; note that these tables can be implemented as TCAM tables. We use "a" as a shorthand for "*accept*" and "d" as a shorthand for "*discard*".

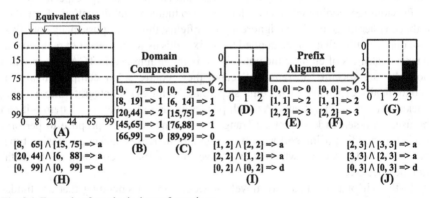

Fig. 8.1 Example of topological transformations

Prefix Alignment: The basic idea of prefix alignment is to "shift", "shrink", or "stretch" ranges by transforming the domain of each field to a new "prefix-friendly" domain so that the majority of the reencoded ranges either are prefixes or can be expressed by a small number of prefixes. In other words, we want to transform a colored hyperrectangle to another one where the ranges align well with prefix boundaries. This will reduce the costs of range expansion. For example, consider the packet classifier in Figure 8.1(I), whose corresponding rectangle is in Figure 8.1(D).

Range expansion will yield 5 prefix rules because interval $[1,2]$ or $[01,10]$ cannot be combined into a prefix. However, by transforming the rectangle in Figure 8.1(D) to the one in Figure 8.1(G), the range expansion of the resulting classifier, as shown in Figure 8.1(J), will have 3 prefix rules because $[2,3]$ is expanded to $1*$. Figures 8.1(D) and (E) show the two transforming tables for F_1 and F_2, respectively. Again, these tables can be implemented in TCAM.

8.1 Topological Transformation

The basic idea of our transformation approach is to transform a given packet classifier into another classifier that can be stored more efficiently in TCAM. Furthermore, we need a transformer that can take any packet and transform it into a new packet that is then used as the search key on the transformed classifier. Of course, the decision that the transformed classifier makes for the transformed packet must be the same as the decision that the original classifier makes for the original packet. We also require that the transformer itself be a packet classifier that can be efficiently stored in TCAM. This is one of the features that differentiates our work from previous reencoding approaches.

More formally, given a d-dimensional packet classifier $\mathbb{C}$ over fields $F_1, \cdots, F_d$, a topological transformation process produces two separate components. The first component is a set of *transformers* $\mathbb{T} = \{\mathbb{T}_i \mid 1 \le i \le d\}$ where transformer $\mathbb{T}_i$ transforms $D(F_i)$ into a new domain $D'(F_i)$. Together, the set of transformers $\mathbb{T}$ transforms the original packet space Σ into a new packet space Σ'. The second component is a transformed d-dimensional classifier $\mathbb{C}'$ over packet space Σ' such that for any packet $(p_1, \cdots, p_d) \in \Sigma$, the following condition holds:

$$\mathbb{C}(p_1, \cdots, p_d) = \mathbb{C}'(\mathbb{T}_1(p_1), \cdots, \mathbb{T}_d(p_d))$$

Each of the d transformers $\mathbb{T}_i$ and the transformed packet classifier $\mathbb{C}'$ are implemented in TCAM.

8.1.1 Architectures

There are two possible architectures for storing the $d+1$ TCAM tables $\mathbb{C}', \mathbb{T}_1, \cdots, \mathbb{T}_d$: the *multi-lookup architecture* and the *pipelined-lookup architecture*, each of which is described below.

In the multi-lookup architecture, which is similar to the multi-lookup architecture in Section 7.1, we store all the $d+1$ tables in one TCAM chip. To identify tables, for each table, we prepend a $\lceil \log(d+1) \rceil$ bit string, which we call the *table ID*, to every entry in the table. Figure 8.2 illustrates the packet classification process using the multi-lookup architecture when $d = 2$. Suppose the three tables are $\mathbb{C}'$, $\mathbb{T}_1$, and

T_2, and their table IDs are 00, 01, and 10, respectively. Given a packet (p_1, p_2), we first concatenate $\mathbb{T}_1$'s table ID 01 with p_1 and use the resulting bit string $01|p_1$ as the search key for the TCAM. Let p_1' denote the search result. Second, we concatenate $\mathbb{T}_2$'s table ID 10 with p_2 and use the resulting bit string $10|p_2$ as the search key for the TCAM. Let p_2' denote the search result. Third, we concatenate the table ID 00 of $\mathbb{C}'$ with p_1' and p_2', and use the resulting bit string $00|p_1'|p_2'$ as the search key for the TCAM. The search result is the final decision for the given packet (p_1, p_2).

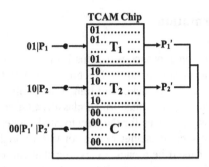

Fig. 8.2 Multi-lookup architecture

We recommend two pipelined-lookup architectures for implementing our transformation approaches: parallel pipelined-lookup and chained pipelined-lookup. In both architectures, we store each of the $d+1$ tables in separate TCAMs. As one TCAM stores only one table, we do not need to prepend table entries with table IDs for either approach. In the parallel pipelined-lookup architecture, the d transformer tables $\mathbb{T}$, laid out in parallel, form a two-element pipeline with the transformed classifier $\mathbb{C}'$. Figure 8.3 illustrates the packet classification process using the parallel pipelined-lookup architecture when $d = 2$. Given a packet (p_1, p_2), we send p_1 and p_2, in parallel over separate buses, to $\mathbb{T}_1$ and $\mathbb{T}_2$, respectively. Then, the search result $p_1'|p_2'$ is used as a key to search on $\mathbb{C}'$. This second search result is the final decision for the given packet (p_1, p_2).

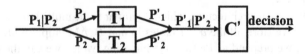

Fig. 8.3 Parallel pipelined-lookup architecture

The $(d+1)$-stage chained pipelined-lookup architecture is similar to the previously proposed pipelined-lookup architecture. Figure 8.4 illustrates the packet classification process using the chained pipelined-lookup architecture when $d = 2$.

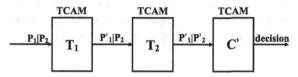

Fig. 8.4 Chained pipelined-lookup architecture

In comparison with the pipelined-lookup architecture, The main advantage of the multi-lookup architecture is that it can be easily deployed since it requires minimal modification of existing TCAM-based packet processing systems. Its main drawback is a modest slowdown in packet processing throughput because $d+1$ TCAM searches are required to process a d-dimensional packet. In contrast, the main advantage of the two pipelined-lookup architectures is high packet processing throughput. Their main drawback is that the hardware needs to be modified to accommodate $d+1$ TCAM chips.

Implementing our reencoding schemes on pipelined-lookup architectures actually improves packet processing throughput over conventional TCAM implementations. While the width of TCAM entries can be set to 36, 72, 144, or 288 bits, the typical TCAM bus width is 72 bits. Thus the conventional TCAM lookup approach, which uses a TCAM entry width of 144 bits, requires either four or five TCAM bus cycles to process a packet: four bus cycles if the decision is stored in TCAM, five bus cycles if the decision is stored in SRAM. Because all the tables produced by our reencoding schemes have width less than 36 bits, we can set TCAM entry width to be 36. Thus, using pipelined-lookup architectures, our reencoding approaches achieve a classification throughput of one packet per cycle; using multi-lookup architectures, our reencoding approaches achieve a classification throughput of one packet per twelve cycles.

8.1.2 Measuring TCAM space

The TCAM space needed by our transformation approach is measured by the total TCAM space needed by the $d+1$ tables: $\mathbb{C}', \mathbb{T}_1, \cdots, \mathbb{T}_d$. We define the space used by a classifier or transformer in a TCAM as the number of entries (i.e., rules) multiplied by the width of the TCAM in bits:

$$space = \# \ of \ entries \times TCAM \ width$$

Although TCAMs can be configured with varying widths, they do not allow arbitrary widths. The width of a TCAM typically can be set at 36, 72, 144, and 288 bits (per entry). The primary goal of the transformation approach is to produce $\mathbb{C}', \mathbb{T}_1, \cdots, \mathbb{T}_d$ such that the TCAM space needed by these $d+1$ TCAM tables is much smaller than the TCAM space needed by the original classifier $\mathbb{C}$. Note that most previous reencoding approaches ignore the space required by the transformers and only focus on the space required by the transformed classifier $\mathbb{C}'$.

8.1.3 TCAM Update

Packet classification rules periodically need to be updated. The common practice for updating rules is to run two TCAMs in tandem where one TCAM is used while the other is updated [Lekkas(2003)]. All our approaches are compatible with this current practice. Because our algorithms are efficient and the resultant TCAM lookup tables are small, updating TCAM tables can be efficiently performed.

If an application requires very frequent rule update (at a frequency less than a second, for example), we can handle such updates in a batch manner by chaining the TCAM chips in our proposed architecture after a TCAM chip of normal width (144 bits), which we call the "hot" TCAM chip. When a new rule comes, we add the rule to the top of the hot TCAM chip. When a packet comes, we first use the packet as the key to search in the hot chip. If the packet has a match in the hot chip, then the decision of the first matching rule is the decision of the packet. Otherwise, we feed the packet to the TCAM chips in our architecture described as above to find the decision for the packet. Although the lookup on the hot TCAM chip adds a constant delay to per packet latency, the throughput can be much improved by pipelining the hot chip with other TCAM chips. Using batch updating, only when the hot chip is about to fill up, we need to run our topological transformation algorithms to recompute the TCAM lookup tables.

8.2 Domain Compression

In this section, we describe our new reencoding scheme called *domain compression*. The basic idea of domain compression is to simplify the logical structure of a classifier by mapping the domain of each field $D(F_i)$ to the smallest possible domain $D'(F_i)$. We formalize this process by showing how a classifier $\mathbb{C}$ defines an equivalence relation on the domain of each of its fields. This equivalence relation allows us to define equivalence classes within each field domain that domain compression will exploit.

Domain compression has several benefits. First, with a compressed domain $D'(F_i)$, we require fewer bits to encode each packet field. This allows us to set TCAM entries widths to be 36 bits rather than 144 bits, which saves both space in the TCAM and time as each entry fits on the 72 bit TCAM bus. Second, each transformed rule r' in classifier $\mathbb{C}'$ will contain fewer equivalence classes than the original rule r did in classifier $\mathbb{C}$. This leads to reduced range expansion and the complete elimination of some rules, which allows us to achieve expansion ratios less than one.

Our domain compression algorithm consists of three steps: (1) computing equivalence classes, (2) constructing transformer $\mathbb{T}_i$ for each field F_i, and (3) constructing the transformed classifier $\mathbb{C}'$.

8.2.1 Step 1: Compute Equivalence Classes

In domain compression, we compress every equivalence class in each domain $D(F_i)$ to a single point in $D'(F_i)$. The crucial tool of the domain compression algorithm is the Firewall Decision Diagram (FDD).

The first step of our domain compression algorithm is to convert a given d-dimensional packet classifier $\mathbb{C}$ to d equivalent reduced FDDs f_1 through f_d where the root of FDD f_i is labeled by field F_i. Figure 8.5(a) shows an example packet classifier over two fields F_1 and F_2 where the domain of each field is [0,63]. Figures 8.5(b) and (c) show the two FDDs f_1 and f_2, respectively. The FDDs f_1 and f_2 are almost reduced except that the terminal nodes are not merged together for illustration purposes.

The crucial observation is that each edge of reduced FDD f_i corresponds to one equivalence class of domain $D(F_i)$. For example, consider the the classifier in Figure 8.5(a) and the corresponding FDD f_1 in Figure 8.5(b). Obviously, for any p_1 and p_1' in $[7,11] \cup [16,19] \cup [39,40] \cup [43,60]$, we have $\mathbb{C}(p_1,p_2) = \mathbb{C}(p_1',p_2)$ for any p_2 in [0,63], so it follows that $\mathbb{C}\{p_1\} = \mathbb{C}\{p_1'\}$.

Theorem 8.1 (Equivalence Class Theorem). *For any packet classifier $\mathbb{C}$ over fields $F_1, \cdots, F_d$ and an equivalent reduced FDD f_i rooted at an F_i node v, the labels of v's outgoing edges are all the equivalence classes over field F_i as defined by $\mathbb{C}$.*

8.2.2 Step 2: Construct Transformers

Given a packet classifier $\mathbb{C}$ over fields $F_1, \cdots, F_d$ and the d equivalent reduced FDDs $f_1, \cdots, f_d$ where the root node of f_i is labeled F_i, we compute transformer $\mathbb{T}_i$ as follows. Let v be the root of f_i, and suppose v has m outgoing edges $e_1, \cdots, e_m$. First, for each edge e_j out of v, as all the ranges in e_j's label belong to the same equivalent class according to Theorem 8.1, we choose one of the ranges in e_j's label to be the representative, which we call the *landmark*. Any of the ranges in e_j's label can be chosen as the landmark. For each equivalence class, we choose the range that intersects the fewest number of rules in $\mathbb{C}$ as the landmark breaking ties arbitrarily. We then sort edges in the increasing order of their landmarks. We use L_j

F_1	F_2	Decision
[12, 15]	[7, 60]	Discard
[41, 42]	[7, 60]	Discard
[20, 38]	[0, 63]	Accept
[0, 63]	[20, 38]	Accept
[7, 60]	[10, 58]	Discard
[1, 63]	[0, 62]	Accept
[0, 62]	[1, 63]	Accept
[0, 63]	[0, 63]	Discard

(a)

⇓

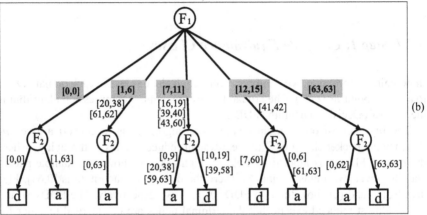

(b)

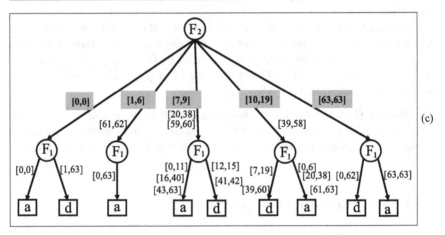

(c)

Fig. 8.5 Step 1 of domain compression

F_1	Decision
$[0,0]$	0
$[1,6] \cup [20,38] \cup [61,62]$	1
$[7,11] \cup [16,19] \cup [39,40] \cup [43,60]$	2
$[12,15] \cup [41,42]$	3
$[63,63]$	4

(a)

F_2	Decision
$[0,0]$	0
$[1,6] \cup [61,62]$	1
$[7,9] \cup [20,38] \cup [59,60]$	2
$[10,19] \cup [39,58]$	3
$[63,63]$	4

(b)

Fig. 8.6 Step 2 of domain compression

F_1	F_2	Decision
$[3,3]$	$[2,3]$	Discard
$\emptyset$	$[2,3]]$	Discard
$\emptyset$	$[0,4]$	Accept
$[0,4]$	$\emptyset$	Accept
$[2,3]$	$[3,3]$	Discard
$[1,4]$	$[0,3]$	Accept
$[0,3]$	$[1,4]$	Accept
$[0,4]$	$[0,4]$	Discard

(a)

⇓

F_1	F_2	Decision
$[3,3]$	$[2,3]$	Discard
$[2,3]$	$[3,3]$	Discard
$[1,4]$	$[0,3]$	Accept
$[0,3]$	$[1,4]$	Accept
$[0,4]$	$[0,4]$	Discard

(b)

Fig. 8.7 Step 3 of domain compression

and e_j to denote the landmark range and corresponding edge in sorted order where edge e_1 has the smallest valued landmark L_1 and edge e_m has the largest valued landmark L_m. Our transformer $\mathbb{T}_i$ then maps all values in e_j's label to value j where $1 \leq j \leq m$. For example, in Figures 8.5(b) and (c), the greyed ranges are chosen as the landmarks of their corresponding equivalence classes, and Figures 8.6(a) and (b) show transformers $\mathbb{T}_1$ and $\mathbb{T}_2$ that result from choosing those landmarks.

8.2.3 Step 3: Construct Transformed Classifier

We now construct transformed classifier $\mathbb{C}'$ from classifier $\mathbb{C}$ using transformers $\mathbb{T}_i$ for $1 \leq i \leq d$ as follows. Let $F_1 \in S_1 \wedge \cdots \wedge F_d \in S_d \rightarrow \langle decision \rangle$ be an original rule in

$\mathbb{C}$. The domain compression algorithm converts $F_i \in S_i$ to $F_i' \in S_i'$ such that for any landmark range L_j $(0 \le j \le m-1)$, $L_j \cap S_i \ne \emptyset$ if and only if $j \in S_i'$. Stated another way, we replace range S_i with range $[a,b] \subseteq D'(F_i)$ where a is the smallest number in $[0, m-1]$ such that $L_a \cap S_i \ne \emptyset$ and b is the largest number in $[0, m-1]$ such that $L_b \cap S_i \ne \emptyset$. Note, it is possible no landmark ranges intersect range S_i; in this case a and b are undefined and $S_i' = \emptyset$. For a converted rule $r' = F_1' \in S_1' \wedge \cdots \wedge F_d' \in S_d' \to \langle decision \rangle$ in $\mathbb{C}'$, if there exists $1 \le i \le d$ such that $S_i' = \emptyset$, then this converted rule r' can be deleted from $\mathbb{C}'$.

For example, consider the rule $F_1 \in [7,60] \wedge F_2 \in [10,58] \to discard$ in the example classifier in Figure 8.5(a). For field F_1, the five landmarks are the five greyed intervals in 8.5(b), namely $[0,0]$, $[1,6]$, $[7,11]$, $[12,15]$, and $[63,63]$. Among these five landmarks, $[7,60]$ overlaps with $[7,11]$ and $[12,15]$, which are mapped to 2 and 3 respectively by transformer $\mathbb{T}_1$. Thus, $F_1 \in [7,60]$ is converted to $F_1' \in [2,3]$. Similarly, for field F_2, $[10,58]$ overlaps with only one of F_2's landmark, $[10,19]$, which is mapped to 3 by F_2's mapping table. Thus, $F_2 \in [10,58]$ is converted to $F_2' \in [3,3]$. Figure 8.7 shows the resultant domain compressed classifier.

Next, we prove that $\mathbb{C}'$ and $\mathbb{T}$ are semantically equivalent to $\mathbb{C}$.

Theorem 8.2. *Consider any classifier $\mathbb{C}$ and the resulting transformers $\mathbb{T}$ and transformed classifier $\mathbb{C}'$. For any packet $p = (p_1, \cdots, p_d)$, we have*
$$\mathbb{C}(p_1, \cdots, p_d) = \mathbb{C}'(\mathbb{T}_1(p_1), \cdots, \mathbb{T}_d(p_d)).$$

Proof. For each field F_i for $1 \le i \le d$, consider p's field value p_i. Let $L(p_i)$ be the landmark range for $\mathbb{C}\{p_i\}$. We set $x_i = \min(L(p_i))$. We now consider the packet $x = (x_1, \cdots x_d)$ and the packets $x(j) = (x_1, \ldots x_{j-1}, p_j, \ldots, p_d)$ for $0 \le j \le d$; that is, in packet $x(j)$, the first j fields are identical to packet x and the last $d-j$ fields are identical to packet p. Note $x(0) = p$ and $x(d) = x$. We now show that $\mathbb{C}(p) = \mathbb{C}(x)$. This follows from $\mathbb{C}(x(0)) = \mathbb{C}(x(1)) = \cdots = \mathbb{C}(x(d))$. Each equality follows from the fact that x_j and p_j belong to the same equivalence class within $D(F_j)$.

Let r be the first rule in $\mathbb{C}$ that packet x matches. We argue that p' will match the transformed rule $r' \in \mathbb{C}'$. Consider the conjunction $F_i \in S_i$ of rule r. Since x matches rule r, it must be the case that $x_i \in S_i$. This implies that $L(p_i) \cap S_i \ne \emptyset$. Thus, by our construction $p_i' = \mathbb{T}_i(p_i) = \mathbb{T}_i(x_i) \in S_i'$. Since this holds for all fields F_i, packet p' matches rule r'. We also argue that packet p' will not match any rule before transformed rule $r' \in \mathbb{C}'$. Suppose packet p' matches some rule $r_1' \in \mathbb{C}'$ that occurs before rule r'. This implies that for each conjunction $F_i \in S_i$ of the corresponding rule $r_1 \in \mathbb{C}$ that $L(p_i) \cap S_i \ne \emptyset$. However, this implies that $x_i \in S_i$ since if any point in $L(p_i)$ is in S_i, then all points in $L(p_i)$ are in S_i. It follows that x matches rule $r_1 \in \mathbb{C}$, contradicting our assumption that rule r was the first rule that x matches in $\mathbb{C}$. Thus, it follows that p' cannot match rule r_1'. It then follows that r' will be the first rule in $\mathbb{C}$ that p' matches and the theorem follows.

8.3 Prefix Alignment

In this section, we describe a new topological transformation approach called *prefix alignment*. When applying this approach, we assume that we have a classifier $\mathbb{C}$ that needs to be converted into a prefix classifier via range expansion. We observe that range explosion happens when ranges do not align well with prefix boundaries. The basic idea of prefix alignment is to "shift", "shrink", or "stretch" ranges by transforming the domain of each field to a new "prefix-friendly" domain so that the majority of the reencoded ranges either are prefixes or can be expressed by a small number of prefixes. This will reduce the costs of range expansion. Of course, we must guarantee that our prefix alignment transformation preserves the semantics of the original classifier.

We first consider the special case where the classifier has only one field F. We develop an optimal solution for this problem using dynamic programming techniques. We then describe how we use this solution as a building block for performing prefix alignment on multi-dimensional classifiers. Finally, we discuss how to compose the two transformations of domain compression and prefix alignment.

8.3.1 Prefix Alignment Overview

The one-dimensional prefix alignment problem can be described as the following "cut" problem. Consider the three ranges $[0, 12]$, $[5, 15]$, and $[0, 15]$ over domain $D(F_1) = [0, 15]$ in classifier $\mathbb{C}$ in Figure 8.8(A), and suppose the transformed domain $D'(F_1) = [00, 11]$ in binary format. Because $D'(F_1)$ has a total of 4 elements, we want to identify three cut points $0 \leq x_1 < x_2 < x_3 \leq 15$ such that if $[0, x_1] \in D(F_1)$ transforms to $00 \in D'(F_1)$, $[x_1 + 1, x_2] \in D(F_1)$ transforms to $01 \in D'(F_1)$, $[x_2 + 1, x_3] \in D(F_1)$ transforms to $10 \in D'(F_1)$, and $[x_3 + 1, 15] \in D(F_1)$ transforms to $01 \in D'(F_1)$, the range expansion of the transformed ranges will have as few rules as possible. For this simple example, there are two families of optimal solutions: those with x_1 anywhere in $[0, 3]$, $x_2 = 4$, and $x_3 = 12$, and those with $x_1 = 4$, $x_2 = 12$, and x_3 anywhere in $[13, 15]$. For the first family of solutions, range $[0, 12]$ is transformed to $[00, 10] = 0 * \cup 10$, range $[5, 15]$ is transformed to $[10, 11] = 1*$, and range $[0, 15]$ is transformed to $[00, 11] = **$. In the second family of solutions, range $[0, 12]$ is transformed to $[00, 01] = 0*$, range $[5, 15]$ is transformed to $[01, 11] = 01 \cup 1*$, and range $[0, 15]$ is transformed to $[00, 11] = **$. The classifier $\mathbb{C}'$ in Figure 8.8(A) shows the three transformed ranges using the first family of solutions. Thus, in both examples, the range expansion of the transformed ranges only has 4 prefix rules while the range expansion of the original ranges has 7 prefix rules.

We now illustrate how to compute an optimal solution using a divide and conquer strategy. The first observation is that we can divide the original problem into two subproblems by choosing the middle cut point. The second observation is that a cut point should be the starting or ending point of a range if possible in order to reduce range expansion. Suppose the target domain $D'(F_1)$ is $[0, 2^b - 1]$. We first need to

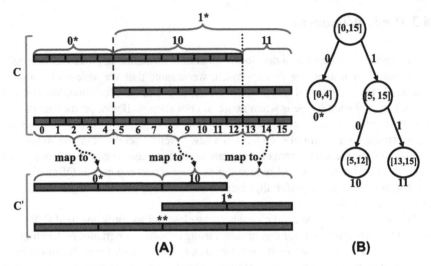

Fig. 8.8 Example of 1-D prefix alignment

choose the middle cut point $x_{2^{b-1}}$, which will divide the problem into two subproblems with target domains $[0, 2^{b-1} - 1] = 0\{*\}^{b-1}$ and $[2^{b-1}, 2^b - 1] = 1\{*\}^{b-1}$ respectively. Consider the example in Figure 8.8(A), the x_2 cut point partitions $[0, 15]$ into $[0, x_2]$, which transforms to prefix $0*$, and $[x_2 + 1, 15]$, which transforms to prefix $1*$. The first observation implies either $x_2 = 4$ or $x_2 = 12$. Suppose we choose $x_2 = 4$; that is, we choose the dashed line as shown in Figure 8.8(A). This then divides the original problem into two subproblems where we need to identify the x_1 cut point in the range $[0, 4]$ and the x_3 cut point in $[5, 15]$. Furthermore, in the two subproblems, we include each range trimmed to fit the restricted domain. For example, in the first subproblem, ranges $[0, 12]$ and $[0, 15]$ are trimmed to $[0, 4]$, and in the second subproblem, ranges $[5, 15]$ and $[0, 15]$ are trimmed to $[5, 15]$ and range $[0, 12]$ is trimmed to $[5, 12]$. It is important to maintain each trimmed range, even though there may be multiple copies of the same trimmed range. We then see in the first subproblem that the choice of x_1 is immaterial since both trimmed ranges span the entire restricted domain. In the second subproblem, the range $[5, 12]$ dictates that $x_3 = 12$ is the right choice.

This divide and conquer process of computing cut points can be represented as a binary cut tree. For example, Figure 8.8(B) depicts the tree where we select $x_2 = 4$ and $x_3 = 12$. This binary cut tree also encodes the transformation from the original domain to the target domain: *all the values in a terminal node will be mapped to the prefix represented by the path from the root to the terminal node.* For example, as the path from the root to the terminal node of $[0, 4]$ is 0, all values in $[0, 4] \in D(F_1)$ are transformed to $0*$.

Note that in the domain compression technique, we considered transformers that mapped points in $D(F_i)$ to points in $D'(F_i)$. In prefix alignment, we consider transformers that map points in $D(F_i)$ to prefix ranges in $D'(F_i)$. If this seems confusing, we can also work with transformers that map points in $D(F_i)$ to points in $D'(F_i)$ with no change in results; however, transformers that map to prefixes more accurately represent the idea of prefix alignment than transformers that map to points. Note also that since we will perform range expansion on $\mathbb{C}'$ with no redundancy removal, we can ignore rule order. We can then view a one-dimensional classifier $\mathbb{C}$ as a multiset of ranges S in $D(F_1)$.

8.3.2 One-dimensional Prefix Alignment

We next present the technical details of our dynamic programming solution to the prefix alignment problem by answering a series of four questions.

8.3.2.1 Correctness of Prefix Alignment

The first question is: why is the prefix alignment transformation process correct? In other words, how does the prefix alignment transformation preserve the semantics of the original classifier? We first define the concept of *prefix transformers* and then show that if prefix transformers are used, the prefix alignment transformation process is correct.

Given a prefix P, we use $\min P$ and $\max P$ to denote the smallest and the largest values in P, respectively.

Definition 8.1 (Prefix transformers). A transformer $\mathbb{T}_i$ is an order-preserving prefix transformer from $D(F_i)$ to $D'(F_i)$ for a packet classifier $\mathbb{C}$ if $\mathbb{T}_i$ satisfies the following three properties. (1) (prefix property) $\forall x \in D(F_i)$, $\mathbb{T}_i(x) = P$ where P is a prefix in domain $D'(F_i)$; (2) (order-preserving property) $\forall x, y \in D(F_i)$, $x < y$ implies either $\mathbb{T}_i(x) = \mathbb{T}_i(y)$ or $\max \mathbb{T}_i(x) < \min \mathbb{T}_i(y)$; (3) (consistency property) $\forall x, y \in D(F_i)$, $\mathbb{T}_i(x) = \mathbb{T}_i(y)$ implies $\mathbb{C}\{x\} = \mathbb{C}\{y\}$.

The following Lemma 8.1 and Theorem 8.3 easily follow the definition of prefix transformers.

Lemma 8.1. *Given any prefix transformer $\mathbb{T}_i$ for a field F_i, for any $a, b, x \in D(F_i)$, $x \in [a, b]$ if and only if $\mathbb{T}_i(x) \in [\min \mathbb{T}_i(a), \max \mathbb{T}_i(b)]$.*

Theorem 8.3 (Topological Alignment Theorem). *Given a packet classifier $\mathbb{C}$ over fields $F_1, \cdots, F_d$, and d prefix transformers $T = \{\mathbb{T}_i \mid 1 \leq i \leq d\}$, and the classifier $\mathbb{C}'$ constructed by replacing any range $[a, b]$ over field F_i ($1 \leq i \leq d$) by the range $[\min \mathbb{T}_i(a), \max \mathbb{T}_i(b)]$, the condition $\mathbb{C}(p_1, \cdots, p_d) = \mathbb{C}'(\mathbb{T}_1(p_1), \cdots, \mathbb{T}_d(p_d))$ holds.*

8.3.2.2 Find Candidate Cut Points

The second question is: what cut points need to be considered? To answer this question, we first introduce the concept of atomic ranges. For any multiset of ranges S (a multiset may have duplicate entries) and any range x over domain $D(F_1)$, we use $S@x$ to denote the set of ranges in S that contain x.

Definition 8.2 (Atomic Range Set). Given a multiset S of ranges, the union of which constitute a range denoted $\bigcup S$, and a set of ranges S', S' is the atomic range set of S if and only if the following four conditions hold: (1) (coverage property) $\bigcup S = \bigcup S'$; (2) (disjoint property) $\forall x, y \in S'$, $x \cap y = \emptyset$; (3) (atomicity property) $\forall x \in S$ and $\forall y \in S'$, $x \cap y \neq \emptyset$ implies $y \subseteq x$; (4) (maximality property) $\forall x, y \in S'$ and $\max x + 1 = \min y$ implies $S@x \neq S@y$.

For any multiset of ranges S, there is one and only one atomic range set of S, which we denote as $AR(S)$. Because of the maximality property of atomic range set, the candidate cut points correspond to the end points of ranges in $AR(S)$. We now show how to compute S-start points and S-end points. For any range $[x, y] \in S$, define the points $x - 1$ and y to be S-end points, and define the points x and $y + 1$ to be S-start points. Note that we ignore $x - 1$ if x is the minimum element of $\bigcup S$; likewise, we ignore $y + 1$ if y is the maximum element of $\bigcup S$. Let $(s_1, \cdots, s_m)$ and $(e_1, \cdots, e_m)$ be the ordered list of S-start points and S-end points. It follows that for $1 \leq i \leq m - 1$ that $s_i \leq e_i = s_{i+1} + 1$. Thus, $AR(S) = \{[s_1, e_1], \cdots, [s_m, e_m]\}$.

For example, if we consider the three ranges in classifier $\mathbb{C}$ in example Figure 8.8(A), range $[0, 12]$ creates S-start point 13 and S-end point 12, range $[5, 15]$ creates S-end point 4 and S-start point 5, and range $[0, 15]$ creates no S-start points or S-end points. Finally, 0 is an S-start point and 15 is an S-end point. This leads to $AR(S) = \{[0, 4], [5, 12], [13, 15]\}$.

8.3.2.3 Choose Target Domain Size

The third question is: how many bits should be used to encode domain $D'(F_1)$? The number of bits b used to encode the domain $D'(F_1)$ may impose some constraints on possible prefix transformers. Consider the example from $\mathbb{C}$ in Figure 8.8(A) with ranges $[0, 12]$, $[5, 15]$, and $[0, 15]$. Suppose there were a fourth range $[5, 7]$. For this multiset of ranges S, we then have $AR(S) = \{[0, 4], [5, 7], [8, 12], [13, 15]\}$. If we allow only 2 bits to encode $D'(F_1)$, then there is only one possible prefix transformer. We must have $[0, 4]$ map to 00, $[5, 7]$ map to 01, $[8, 12]$ map to 10, and $[13, 15]$ map to 11. On the other hand, if we allow 3 bits, we can also allow additional prefix transformers such as $[0, 4]$ map to 000, $[5, 7]$ map to 001, $[8, 12]$ map to 01*, and $[13, 15]$ map to 1**. In this case, the first prefix transformer is superior to this second prefix transformer. However, if the original ranges had been $[0, 4]$, $[0, 7]$, $[0, 12]$, and $[0, 15]$, the second prefix transformer would have been superior, and this prefix transformer is only possible if we encode $D'(F_1)$ with at least 3 bits.

We will include the number of bits b used to encode $D'(F_1)$ as an input parameter to our prefix alignment problem. We determine the best b through an iterative process of repeatedly incrementing b and computing an optimal solution for that b. We start by choosing $b = \lceil \log_2 |AR(S)| \rceil$, which is the smallest possible number of bits for any legal prefix transformer. Once we have a solution, we increment b and repeat the process until we cannot reduce the range expansion any further. We choose a linear search as opposed to a binary search for efficiency reasons. As we shall see in a moment, any solution using b bits will require a sub-solution using $b-1$ bits. Thus, when we fail to find a solution using b bits and try to find a solution using $2b$ bits, we will require a sub-solution for each number from $b+1$ to $2b-1$ (otherwise we would have found a solution using b bits). Furthermore, the binary search may miss the best b by a large factor, which will lead to a large amount of unnecessary computation.

8.3.2.4 Choose Optimal Cut Points

The fourth question is: How do we choose the optimal cut points? As we noted before, we can view a one-dimensional classifier $\mathbb{C}$ as a multiset of ranges S in $D(F_1)$. We then formulate the one-dimensional prefix alignment problem as follows: *Given a multiset of ranges S over field F_1 and a number of bits b, find prefix transformer $\mathbb{T}_1$ such that the range expansion of the transformed multiset of ranges S' has the minimum number of prefix rules and $D'(F_1)$ can be encoded using only b bits.*

We now present an optimal solution for this problem using dynamic programming. Given a multiset of ranges S, we first compute its atomic range set $AR(S)$. Suppose there are m atomic ranges $R_1, \cdots, R_m$ with S-start points s_1 through s_m and S-end points e_1 through e_m sorted in increasing order. For any S-start point s_x and S-end point s_y where $1 \le x \le y \le m$, we define $S \cap [x,y]$ to be the multiset of ranges from S that intersect range $[s_x, s_y]$; furthermore, we assume that each range in $S \cap [x,y]$ is trimmed so that its start point is at least s_x and its end point is at most s_y. We then define a collection of subproblems as follows. For every $1 \le x \le y \le m$, we define a prefix alignment problem $PA(x,y,b)$ where the problem is to find a prefix transformer $\mathbb{T}_1$ for $[s_x, e_y] \subseteq D(F_1)$ such that the range expansion of $(S \cap [x,y])'$ has the smallest possible number of prefix rules and the transformed domain $D'(F_1)$ can be encoded in b bits. We use $cost(x,y,b)$ to denote the number of prefix rules in the range expansion of the optimal $(S \cap [x,y])'$. The original prefix alignment problem then corresponds to $PA(1,m,b)$ where b can be arbitrarily large.

The key observation that allows the use of dynamic programming is that the prefix alignment problem obeys the optimal substructure property. For example, consider $PA(1, m, b)$. As we employ the divide and conquer strategy to locate a middle cut point that will establish what the prefixes $0\{*\}^{b-1}$ and $1\{*\}^{b-1}$ correspond to, there are $m-1$ choices of cut points to consider: namely e_1 through e_{m-1}. Suppose the optimal cut point is e_k where $1 \le k \le m-1$. Then the optimal solution to $PA(1,m,b)$ will build upon the optimal solutions to subproblems $PA(1,k,b-1)$ and $PA(k+1,m,b-1)$. That is, the optimal prefix transformer for

$PA(1,m,b)$ will simply append a 0 to the start of all prefixes in the optimal prefix transformer for $PA(1,k,b-1)$, and similarly it will append a 1 to the start of all prefixes in the optimal prefix transformer for $PA(k+1,m,b-1)$. Moreover, $cost(1,m,b) = cost(1,k,b-1) + cost(k+1,m,b-1) - |S@[1,m]|$. We subtract $|S@[1,m]|$ in the above cost equation because ranges that include all of $[s_1,e_m]$ are counted twice, once in $cost(1,k,b-1)$ and once in $cost(k+1,m,b-1)$. However, as $[s_1,e_k]$ transforms to $0\{*\}^{b-1}$ and $[s_{k+1},e_m]$ transforms to $1\{*\}^{b-1}$, the range $[s_1,e_m]$ can be expressed by one prefix $\{*\}^b = 0\{*\}^{b-1} \cup 1\{*\}^{b-1}$.

Based on this analysis, Theorem 8.4 shows how to compute the optimal cuts and binary tree. As stated earlier, the optimal prefix transformer $\mathbb{T}_1$ can then be computed from the binary cut tree.

Theorem 8.4. *Given a multiset of ranges S with $|AR(S)| = m$, $cost(l,r,b)$ for any $b \geq 0, 1 \leq l \leq r \leq m$ can be computed as follows. For any $1 \leq l < r \leq m$, and $1 \leq k \leq m$, and $b \geq 0$:*

$$cost(l,r,0) = \infty,$$
$$cost(k,k,b) = |S@[k,k]|,$$

and for any $1 \leq l < r \leq m$ and $b \geq 1$

$$cost(l,r,b) = \min_{k \in \{l,\dots,r-1\}} \begin{pmatrix} cost(l,k,b-1) \\ + \\ cost(k+1,r,b-1) \\ - \\ |I@[l,r]| \end{pmatrix}$$

□

Note that we set $cost(k,k,0)$ to $|S@[k,k]|$ for the convenience of the recursive case. The interpretation is that with a 0-bit domain, we can allow only a single value in $D'(F_1)$; this single value is sufficient to encode the transformation of an atomic interval.

8.3.3 Multi-Dimensional Prefix Alignment

We now consider the multi-dimensional prefix alignment problem. Unfortunately, while we can optimally solve the one-dimensional problem, there are complex interactions between the dimensions that make solving the multi-dimensional problem optimally extremely difficult. In particular, the total range expansion required for each rule is the product of the range expansion required for each field. Thus, there may be complex tradeoffs where we sacrifice one field of a rule but align another field so that the costs do not multiply. However, we have not found a polynomial algorithm for optimally choosing which rules to align well in which fields. It is an open problem to prove whether the optimal multi-dimensional prefix alignment problem is NP-hard.

In this chapter, we present a hill-climbing solution where we iteratively apply our one-dimensional prefix alignment algorithm one field at a time to improve our solution. The basic idea is to perform prefix alignment one field at a time; however, because the range expansion of one field affects the numbers of ranges that appear in the other fields, we run prefix alignment for each field more than once. Running prefix alignment more than once allows each field to use more and more accurate information about the number of times each range appears in a field.

For a classifier $\mathbb{C}$ over fields $F_1, \ldots, F_d$, we first create d identity prefix transformers $\mathbb{T}_1^0, \ldots, \mathbb{T}_d^0$. We define a *multi-field prefix alignment iteration k* as follows. For i from 1 to d, generate the optimal prefix transformer $\mathbb{T}_i^k$ assuming the prefix transformers for the other fields are $\{\mathbb{T}_1^{k-1}, \ldots, \mathbb{T}_{i-1}^{k-1}, T_{i+1}^{k-1}, \ldots, T_d^{k-1}\}$. Our iterative solution starts at $k = 1$ and preforms successive multi-field prefix alignment iterations until no improvement is found for any field.

8.3.4 Composing with Domain Compression

Although the two transformation approaches proposed in this chapter can be used individually to save TCAM space, we advocate combining them together to achieve higher TCAM reduction. Given a classifier $\mathbb{C}$ over fields $F_1, \ldots, F_d$, we first perform domain compression resulting in a transformed classifier $\mathbb{C}'$ and d transformers $\mathbb{T}_1^{dc}, \ldots, \mathbb{T}_d^{dc}$; then, we perform prefix alignment on the classifier $\mathbb{C}'$ resulting in a transformed classifier $\mathbb{C}''$ and d transformers $\mathbb{T}_1^{pa}, \ldots, \mathbb{T}_d^{pa}$. To combine the two transformation processes into one, we merge each pair of transformers $\mathbb{T}_i^{dc}$ and $\mathbb{T}_i^{pa}$ into one transformer $\mathbb{T}_i$ for $1 \le i \le d$. One nice property of their composition is that since the transformer for domain compression is a function from $D(F_i)$ to a point in $D'(F_i)$ and each point in $D'(F_i)$ will belong to its own equivalence class in $D'(F_i)$ for $1 \le i \le d$, each point $x \in D'(F_i)$ defines an atomic range $[x, x]$.

A good property of the two proposed topological transformation approaches is that they are composable with many other reencoding or TCAM optimization techniques. For example, we can apply previous TCAM minimization schemes (such as [Liu and Gouda(2005), Dong et al(2006)Dong, Banerjee, Wang, Agrawal, and Shukla, Meiners et al(2007)Meiners, Liu, and Torng]) to a transformed classifier to further reduce TCAM space. Furthermore, as new classifier minimization algorithms are developed, our transformations can potentially leverage these future results. Finally, we can apply the optimal algorithm in [Suri et al(2003)Suri, Sandholm, and Warkhede] to compute the minimum possible transformers $\mathbb{T}_i$ for $1 \le i \le d$.

References

[Applegate et al(2007)Applegate, Calinescu, Johnson, Karloff, Ligett, and Wang] Applegate DA, Calinescu G, Johnson DS, Karloff H, Ligett K, Wang J (2007) Compressing rectilinear pictures and minimizing access control lists. In: Proc. ACM-SIAM Symposium on Discrete Algorithms (SODA)

[Baboescu et al(2003)Baboescu, Singh, and Varghese] Baboescu F, Singh S, Varghese G (2003) Packet classification for core routers: Is there an alternative to CAMs? In: Proc. IEEE INFOCOM

[Bremler-Barr and Hendler(2007)] Bremler-Barr A, Hendler D (2007) Space-efficient TCAM-based classification using gray coding. In: Proc. 26th Annual IEEE conf. on Computer Communications (Infocom)

[Che et al(2008)Che, Wang, Zheng, and Liu] Che H, Wang Z, Zheng K, Liu B (2008) DRES: Dynamic range encoding scheme for tcam coprocessors. IEEE Transactions on Computers 57(7):902–915

[Dong et al(2006)Dong, Banerjee, Wang, Agrawal, and Shukla] Dong Q, Banerjee S, Wang J, Agrawal D, Shukla A (2006) Packet classifiers in ternary CAMs can be smaller. In: Proc. ACM Sigmetrics, pp 311–322

[Draves et al(1999)Draves, King, Venkatachary, and Zill] Draves R, King C, Venkatachary S, Zill B (1999) Constructing optimal IP routing tables. In: Proc. IEEE INFOCOM, pp 88–97

[Eastlake and Jones(2001)] Eastlake D, Jones P (2001) Us secure hash algorithm 1 (sha1). RFC 3174

[Feldmann and Muthukrishnan(2000)] Feldmann A, Muthukrishnan S (2000) Tradeoffs for packet classification. In: Proc. 19th IEEE INFOCOM, URL citeseer.nj.nec.com/feldmann00tradeoffs.html

[Garey and Johnson(1978)] Garey MR, Johnson DS (1978) Strong NP-completeness results: motivation, examples, and implications. Journal of ACM 25(3):499–508

[Gouda and Liu(2004)] Gouda MG, Liu AX (2004) Firewall design: consistency, completeness and compactness. In: Proc. 24th IEEE Int. conf. on Distributed Computing Systems (ICDCS-04), pp 320–327, URL http://www.cs.utexas.edu/users/alex/publications/fdd.pdf

[Gouda and Liu(2007)] Gouda MG, Liu AX (2007) Structured firewall design. Computer Networks Journal (Elsevier) 51(4):1106–1120

[Gupta and McKeown(1999a)] Gupta P, McKeown N (1999a) Packet classification on multiple fields. In: Proc. ACM SIGCOMM, pp 147–160, URL citeseer.nj.nec.com/gupta99packet.html

[Gupta and McKeown(1999b)] Gupta P, McKeown N (1999b) Packet classification using hierarchical intelligent cuttings. In: Proc. Hot Interconnects VII

[Gupta and McKeown(2001)] Gupta P, McKeown N (2001) Algorithms for packet classification. IEEE Network 15(2):24–32, URL citeseer.nj.nec.com/article/gupta01algorithms.html

[Hamming(1950)] Hamming RW (1950) Error detecting and correcting codes. Bell Systems Technical Journal 29:147–160

[Lakshman and Stiliadis(1998)] Lakshman TV, Stiliadis D (1998) High-speed policy-based packet forwarding using efficient multi-dimensional range matching. In: Proc. ACM SIGCOMM, pp 203–214, URL citeseer.nj.nec.com/lakshman98highspeed.html

[Lakshminarayanan et al(2005)Lakshminarayanan, Rangarajan, and Venkatachary] Lakshminarayanan K, Rangarajan A, Venkatachary S (2005) Algorithms for advanced packet classification with ternary CAMs. In: Proc. ACM SIGCOMM, pp 193 – 204

[Lekkas(2003)] Lekkas PC (2003) Network Processors - Architectures, Protocols, and Platforms. McGraw-Hill

[Liu and Gouda(2004)] Liu AX, Gouda MG (2004) Diverse firewall design. In: Proc. Int. conf. on Dependable Systems and Networks (DSN-04), pp 595–604

[Liu and Gouda(2005)] Liu AX, Gouda MG (2005) Complete redundancy detection in firewalls. In: Proc. 19th Annual IFIP conf. on Data and Applications Security, LNCS 3654, pp 196–209, URL http://www.cs.utexas.edu/users/alex/publications/Redundancy/redundancy.pdf

[Liu and Gouda(to appear)] Liu AX, Gouda MG (to appear) Complete redundancy removal for packet classifiers in tcams. IEEE Transactions on Parallel and Distributed Systems (TPDS)

[Liu et al(2008)Liu, Meiners, and Zhou] Liu AX, Meiners CR, Zhou Y (2008) All-match based complete redundancy removal for packet classifiers in TCAMs. In: Proc. 27th Annual IEEE conf. on Computer Communications (Infocom)

[Liu(2002)] Liu H (2002) Efficient mapping of range classifier into Ternary-CAM. In: Proc. Hot Interconnects, pp 95– 100

[van Lunteren and Engbersen(2003)] van Lunteren J, Engbersen T (2003) Fast and scalable packet classification. IEEE Journals on Selected Areas in Communications 21(4):560– 571

[McGeer and Yalagandula(2009)] McGeer R, Yalagandula P (2009) Minimizing rulesets for tcam implementation. In: Proc. IEEE Infocom

[Meiners et al(2007)Meiners, Liu, and Torng] Meiners CR, Liu AX, Torng E (2007) TCAM Razor: A systematic approach towards minimizing packet classifiers in TCAMs. In: Proc. 15th IEEE conf. on Network Protocols (ICNP), pp 266–275, Reprinted, with permission

[Meiners et al(2008a)Meiners, Liu, and Torng] Meiners CR, Liu AX, Torng E (2008a) Algorithmic approaches to redesigning tcam-based systems [extended abstract]. In: Proc. ACM SIGMETRICS

[Meiners et al(2008b)Meiners, Liu, and Torng] Meiners CR, Liu AX, Torng E (2008b) Bit weaving: A non-prefix approach to compressing packet classifiers in tcams. In: Proc. IEEE ICNP, Reprinted with permission

[Meiners et al(2008c)Meiners, Liu, and Torng] Meiners CR, Liu AX, Torng E (2008c) Topological transformation approaches to optimizing tcam-based packet processing systems. In: Proc. ACM SIGCOMM (poster session)

[Pao et al(2006)Pao, Li, and Zhou] Pao D, Li YK, Zhou P (2006) An encoding scheme for TCAM-based packet classification. In: Proc. 8th IEEE Int. conf. on Advanced Communication Technology (ICACT)

[Pao et al(2007)Pao, Zhou, Liu, and Zhang] Pao D, Zhou P, Liu B, Zhang X (2007) Enhanced prefix inclusion coding filter-encoding algorithm for packet classification with ternary content addressable memory. Computers & Digital Techniques, IET 1:572–580

[Qiu et al(2001)Qiu, Varghese, and Suri] Qiu L, Varghese G, Suri S (2001) Fast firewall implementations for software-based and hardware-based routers. In: Proc. the 9th Int. conf. on Network Protocols (ICNP), URL citeseer.nj.nec.com/qiu01fast.html

[Rivest(1992)] Rivest R (1992) The md5 message-digest algorithm. RFC 1321

[iva san et al(1999)iva san, Suri, and Varghese] iva san VS, Suri S, Varghese G (1999) Packet classification using tuple space search. In: Proc. ACM SIGCOMM, pp 135–146, URL citeseer.nj.nec.com/srinivasan99packet.html

[boe scu and Varghese(2001)] boe scu FB, Varghese G (2001) Scalable packet classification. In: Proc. ACM SIGCOMM, pp 199–210, URL http://citeseer.nj.nec.com/baboescu01scalable.html

[Singh et al(2003)Singh, Baboescu, Varghese, and Wang] Singh S, Baboescu F, Varghese G, Wang J (2003) Packet classification using multidimensional cutting. In: Proc. ACM SIGCOMM, pp 213–224, URL http://www.cs.ucsd.edu/~varghese/PAPERS/hyp-sigcomm03.pdf

[Spitznagel et al(2003)Spitznagel, Taylor, and Turner] Spitznagel E, Taylor D, Turner J (2003) Packet classification using extended TCAMs. In: Proc. 11th IEEE Int. conf. on Network Protocols (ICNP), pp 120–131

[Srinivasan et al(1998)Srinivasan, Varghese, Suri, and Waldvogel] Srinivasan V, Varghese G, Suri S, Waldvogel M (1998) Fast and scalable layer four switching. In: Proc. ACM SIGCOMM, pp 191–202, URL citeseer.nj.nec.com/srinivasan98fast.html

[Suri et al(2003)Suri, Sandholm, and Warkhede] Suri S, Sandholm T, Warkhede P (2003) Compressing two-dimensional routing tables. Algorithmica 35:287–300

[Taylor and Turner(2005)] Taylor D, Turner J (2005) Scalable packet classification using distributed crossproducting of field labels. In: Proc. IEEE INFOCOM, pp 269–280

[Taylor(2005)] Taylor DE (2005) Survey & taxonomy of packet classification techniques. ACM Computing Surveys 37(3):238–275

[Woo(2000)] Woo TYC (2000) A modular approach to packet classification: Algorithms and results. In: Proc. IEEE INFOCOM, pp 1213–1222, URL citeseer.nj.nec.com/woo00modular.html

[Yu et al(2005)Yu, Lakshman, Motoyama, and Katz] Yu F, Lakshman TV, Motoyama MA, Katz RH (2005) SSA: A power and memory efficient scheme to multi-match packet classification. In: Proc. Symposium on Architectures for Networking and Communications Systems (ANCS), pp 105–113

Index